PAL Publications has become the leader in providing the industrial trades with valuable, time saving information. Our books are manufactured to the highest quality standards in the industry. All orders are processed in-house and shipped the next business day. We stand behind every book sold by providing exceptional service and customer satisfaction.

OTHER BOOKS BY JOHN GLADSTONE

AIR CONDITIONING & MECHANICAL TRADES

Written for those preparing for a Contractors License Exam, this book reveals tactics used by the author including previous test questions and answers covering air conditioning, refrigeration, heating, sheet metal, insulation, piping and service.
(442 pages)

JOURNEYMAN GENERAL MECHANICAL EXAMINATION

An illustrated review designed to prepare practicing A/C contractors and senior year apprentices for the journeyman's HVAC or General Mechanical Exam. Includes over 500 questions and answers from previous municipal and state certification tests. (265 pages)

CONTRACTORS EXAM BOOK

Designed and written to help candidates prepare for the certified General and Residential Contractor Licensing Exams. Contains valuable information on site plans, excavation, framework, concrete, masonry, framing, steel construction and much more.
(243 pages)

AIR CONDITIONING TESTING, ADJUSTING AND BALANCING

A field practice manual for the contractor and serviceman containing charts, tables, formulas and laws along with basic knowledge on the latest techniques for fine tuning HVAC systems. (196 pages)

MOVING AIR THROUGH FANS & DUCTS

An extensive coverage including multi-rating fan tables, non-standard air calculations, pressure equivalents and conversion data, fan and duct calculations as well as a study guide for contractor and journeymae license exams. (87 pages)

MOVING WATER THROUGH PUMPS & PIPES FOR HPAC

Subjects including general hydraulics, pump and cooling tower selection, sizing pipe and fittings, bending pipe and calculating offsets. A basic guide to effectively move fluids. (88 pages)

DUCTWORK ESTIMATING FOR HVAC

Created to improve an estimator's accuracy while saving time, the information provided encompasses sheetmetal production and weight tables, charts, formulas and specifications necessary in a quick reference format. (42 pages)

JOURNEYMAN GENERAL MECHANICAL EXAMINATION:

An Illustrated Review with Questions and Answers

Second Edition

by John Gladstone, B.A., M.A.

Life Member, American Society of Heating, Refrigerating and Air-Conditioning Engineers; Certificate Member, Refrigeration Service Engineers Society; Former Chairman of the Examinations Committee of the Dade County Mechanical Contractors Examining Board; Founding Member, Associated Air Balance Council (Retired).

AIR CONDITIONING • REFRIGERATION • PIPING • HEATING • RIGGING • GENERAL SAFETY • MATHEMATICS • FORMULAS • TABLES • CHARTS

JOURNEYMAN GENERAL
MECHANICAL EXAMINATION:
AN ILLUSTRATED REVIEW WITH QUESTIONS AND ANSWERS

SECOND EDITION

ISBN: 0-930644-11-5

1st printing, 1973
2nd printing, 1976
3rd printing, paperback 1979
4th printing, expanded edition, 1982
5th printing, expanded edition, 1982
6th printing, expanded edition, 1982
7th printing, second edition, 1988
8th printing, 1992
9th printing, with revisions, 1995
10th printing, 2000

Library of Congress Catalog Card Number 88-0804593

PAL PUBLICATIONS, 374 Circle of Progress, Pottstown, PA 19464
Voice (800) 246-2175 Fax (800) 396-1663

Contents

PART 5 ANSWERS TO QUIZZES

PART 6 APPENDIX

PREFACE TO THE SECOND EDITION

The principle aim of this book is to prepare practicing journeymen and senior-year apprentices for the journeymen's air conditioning and general mechanical examination; to help him or her pass any municipal or state examination for certification. It is, therefore, a practical book--a fact book.

Although many schools for air conditioning and refrigeration are in place today, the industry is crying out desperately for skilled, well-trained mechanics. In checking the number of school graduates who have successfully passed a journeymen's general mechanical examination, the author has found that only a tiny fraction of the people completing a school course of any kind were able to pass a municipal certificate examination without further study. An exception to this pattern is the graduate from the Joint Apprenticeship Schools administered by the United Association of Journeymen and Apprentices of the Plumbing and Pipefitting Industry,. As far as this author is aware no schools in the nation can compare with the UA training program for Journeyman General Examination.

The number of air conditioning courses offered in vocational and trade schools throughout the country has proliferated. High schools, colleges, and adult extension courses provide various levels of training, and these are supplemented by a growing number of private technical schools offering either regular classroom training or home study correspondence courses. But the journeyman general mechanical must know much more. Albeit different states and municipalities vary in their certification and licensing programs, usually, a journeyman who works for a mechanical contractor must be qualified journeyman general mechanical; that is, he must be a qualified to perform in air conditioning, refrigeration, heating, and pressure piping trades. And in some areas he must also be qualified to do sheet metal, ventilation and insulation. He must be knowledgeable in gas cooling as well as gas heating, large ammonia systems as well as centrifugals, construction, service and test and balance.

There are many excellent air conditioning mechanics who have little knowledge of pipefitting, many skilled service men who cannot erect a 50-ton cooling tower, many good heating men unable to perform the test/adjust/balance function on a total environmental control system, many good sheet metal men who can erect an air handler but know nothing about controls.

It is intended through this book to give the exam candidate the kind of information needed in the examination room but not generally found in other books. The examination is not a practical field test, it is test of your knowledge and understanding of air conditioning principles, pipefitting, ductwork, service, installation, safety, rigging and certain concepts of the air conditioning code. It is also an indication of your literacy and ability to read and study.

This book will put you on the right path. It will teach you how to handle certain concepts, how to respond to examination problems, and what to expect in the exam room. It will get you ready.

In the fifteen years since the first edition appeared, thousands of men and women have used this book to help them pass their exams and many schools have used it as a text. We hope that this revised 2nd edition will be as beneficial to as many people and will not require revision for another fifteen years.

Much source material was necessary to prepare this book. For providing various material--copyrighted and otherwise--I wish to thank the following organizations: the American Society of Heating, Refrigerating and Air Conditioning Engineers; American Society of Mechanical Engineers; Copper Development Association; National Fire Protection Association; Air Conditioning National Association; Sheet Metal and Air Conditioning National Association, Manufacturer's Standardization Society of the Valve and Fittings Industry; Allis-Chalmers Manufacturing Company; Bacharach Instrument Company; The Crane Company; Mueller Brass Company; Mechanical Power transmission Association; Henry Valve Company; Ric-Wil Company; Rubber Manufacturer's Association and Spirax Sarco Inc.

John Gladstone
Coral Gables, Fl

INTRODUCTION

Unfortunately, there are no existing "Standards" for writing examinations, testing, or grading at this time. Neither are there any standards set for experience, residency or educational qualifications for applicants. Examinations and qualification requirements vary from place to place. But the basics do not change, most exams will ask questions about service and construction. Some questions will be practical, some theoretical. In addition most examinations will test the candidates knowledge on Code Questions.

Although different areas around the country use different building codes, most local codes are usually referral codes, that is, they refer to other standards and are based, in general, upon those standards. For this reason, most codes are quite similar although, of course, there may be sharp regional differences owing to hurricanes, earthquakes and other conditions that are unique to specific areas. Such standards as ASHRAE 15 Safety Code for Mechanical Refrigeration; ASME B31.9 piping code; SMACNA duct construction and installation manuals and the NFPA National Fire Code Standards make up the basis for all local codes. You should be thoroughly familiar with these standards.

Because South Florida was the first to require certification examinations for journeyman general mechanical and many other regions around the country have borrowed from these exams through the years, many of the questions and problems in this book have been taken for those exams. Many of these examination questions were written by me twenty-five years ago when I was Chairman of the Mechanical Contractor's Examining Board and are still current.

This book is divided into six parts. In **Part 1** I have tried to familiarize the reader with the language and construction of tests generally so he need not walk into the exam room cold. To many, the thought of sitting in a "classroom" or "examination room" is a disturbing and intimidating experience; I have given some hints about what to expect in the exam room and hints about what, where, when and how to study. **Part 1** ends with 300 of the most frequently asked questions--they should be reviewed over and over and over again.

It is suggested that you treat the questions found in the following pages as an actual exam. Before looking for answers, find a quiet place to sit down, and work the questions as though you were in an actual examination room. Allow an average of 2-½ minutes per question. Of course, some questions will be much more difficult than others, and some will be problem type questions, but on average, you will have to answer 25 questions in an hour to pass the exam. Examiners usually allow 4 hours for a 100 questions exam or 8 hours for a 200 questions exam; to complete your work, you will have to answer each question in approximately 2-½ minutes. That will require a good deal of concentration and self-discipline. Timing is essential in any exam; do not spend more time on one question than is allowable. If you can't answer a question in the allotted time, pass it by and go to the next, returning afterwards if you have additional time left at the remainder of the exam.

After you have answered all the 300 questions in the allowable time, check your answers with those at the end of each section and mark your grade. The passing grade is usually 70%. Allowing 0.333 points per each correct answer, you will have had to answer 210 correctly to achieve a passing grade. If you examine the questions you missed, you will notice that they fall into some kind of pattern; your weakness will either be in a particular category such as heating or pressure piping, or you will not be sharp enough with formulas, or what-have-you. Whatever the case may be, your weakness is the area that should receive your concentrated attention and study.

Part 2 is the technical section beginning with a basic math review and ending with symbols; very important for recognition tests. All the basic formulas and tables are presented here. I have tried to keep this material lean, and have relied more on modern graphic illustrations than on complex calculations. Charts, tables and "Speed-0-Graphs" are used to help the reader find answers quickly; time is the most important element in any examination.

Part 3 deals with basic field application problems. Concentrate on those subjects that are your weakness. In those areas in which you are particularly weak go to your other reference books for in-depth study. This book is designed to give you the confidence and experience you need to take, and pass any journeyman's certification exam. It is not designed to teach you your trade or profession.

Part 4 offers quizzes and tests in specific subject areas related to the information presented in Part 2, and Part 3 and concludes with a typical "final exam." Our strategy is to direct your studies in the right places, help structure your study plans, show you how to study and provide quizzes and tests to sharpen your test-taking skills. The questions and quizzes presented here are actual examinations and are representative of what one might expect to encounter in the exam room.

Part 5 gives the answers to all quizzes and tests as well as to the 300 Questions in Part 1.

Part 6 is the Appendix, it represents important conversion tables and other quick-reference material.

PART 1

EVERYTHING FOR THE EXAM

HOW TO USE THIS STUDY COURSE

The course of study is carefully organized to develop your facilities for one purpose; passing the license examination. We might draw an analogy of a boxer training for his championship bout. He must conform to the discipline and organization imposed upon him by his manager and trainer, everything else in his life is pushed aside until after the big fight. Much of your success the day of the examination, depends on your ability to organize *now*. Do not skip through this course or hop about from one part to another. Follow the plan of organization as it is presented and understand each instruction before proceeding to the next.

You're a busy man--don't waste time. Make a study schedule and stick to it! The *STUDY SCHEDULE*, Figure 1.1, is designed to help you get organized. Fill in the maximum allowable study time you can afford. Investigation has proven that frequent short intervals of study at regular periods is more rewarding than less frequent but longer periods. However, this might not always be possible. Keep your schedule realistic . If your schedule is too ambitious and you can't meet it, it will break down your organization and you will lose the potential value of the course.

Assume you have marked your *STUDY SCHEDULE* to allow 1 hour of study each night plus 4 hours on Sunday. After the second week of study you discover that you have a conflict--owing to an association meeting or union meeting--on the second Tuesday of each month. It would then be wise to drop Tuesday out of the schedule completely...make that your night off each week and try to fit in other regular activities for the remaining Tuesdays, such as taking your wife to a movie. If your first plan does not work out well enough for you, change it to suit your particular circumstances, then stay rigidly with your new plan.

The examination announcement will usually list the reference material required for your particular exam. Certain books and pamphlets must be sent for by mail and are slow arriving. Do not procrastinate; get your necessary books and pamphlets as soon as possible. Remember, your profession requires certain reference material on your book shelf at all times. Money spent for this purpose is well spent, an investment you will never regret. Sometimes, one fact culled from a $25.00 book can save you hundreds, or even thousands of dollars on the job. Your accountant will advise you that expenditures for professional books are deductible from your income taxes.

THE 10 RULES FOR EFFICIENT STUDY

1. Make a schedule and stick to it. It will raise your level of personal efficiency. It will diminish emotional strain and lighten the burden. It will help you to master concentration. It will organize your entire family and reduce their interference with your program.

2. Study in the same quiet and well lighted place each time.

3. Keep the top of your desk or study table clear of all unrelated material. Do not wander off your course.

4. Start each study period by the clock, promptly, and end it the same way.

5. If your study sessions are long, take short rests periodically to relieve tension and stiffness.

6. Study is not reading: As you study, evaluate what you are trying to learn. Why is this expressed this way? What is it for? Can it be done another way?

7. Keep a pencil in your hand while you are studying, and a ruled pad or notebook alongside. A difficult or important passage should be written out; such an expression will help plant the thought firmly in your mind. Summarize ideas in the margin , underscore important, passages, rework hard to grasp ideas.

8. Keep a good dictionary of the English language on your desk. Be sure you know the meanings of all the words.

9. Don't get up from your work until it is time. If you need to have a smoke or nibble some pretzels with beer to keep relaxed during your study period, have these things ready *before* you start work.

10. Review your work constantly. Never, never, pass up a review because you feel you have already mastered that lesson.

S T U D Y   S C H E D U L E

This is your work plan; give it your meticulous attention. Develop a realistic schedule of the most possible hours you can devote to this program — then stick to it with an iron discipline! Draw an X through the hours you intend to set aside for study under each day and post this schedule in a visible place for constant reference.

	SUNDAY	MONDAY	TUESDAY	WEDNESDAY	THURSDAY	FRIDAY	SATURDAY
7:00							
8:00							
9:00							
10:00							
11:00							
12:00							
1:00							
2:00							
3:00							
4:00							
5:00							
6:00							
7:00							
8:00							
9:00							
10:00							
11:00							

FIGURE 1.1

PREPARING FOR THE EXAMINATION

An examination may be either "open book" or "closed book" or a combination of both. An "open book" exam means that the candidate is expected to solve problems and answer questions by referring to published books, other data, or notes. This is a pratical type exam and is usually based on given conditions which you are likely to encounter any time on an actual job. The "closed book" exam, or portion of exam, prohibits the use of any reference material in the test room. This is also, generally, a practical exam, but you may occasionally see some questions that are impractically presented. For example, you may be asked--on a refrigeration master exam--to give the operating pressure of Ref-717 at 290°F. Or, you may be asked--on an electrical master exam--to give the minimum required meter room dimensions for housing 23 meters. These examples of "closed book" memory type questions are seldom encountered; usually a memory type question will deal with a fundamental law or formula, something you work with every day. Do not attempt to memorize tables, charts or any questions and answers in this course. Although some of the questions you will see may reappear on your actual examination, it would be very unwise to attempt to memorize.

The "open book" section of the examination will really be a test of your ability to use available reference material, that is, you will be required to dig out certain information in a limited time. Time is the most critical thing about any examination, and the better organized you are to use your time properly, the better your chance to pass. Familiarize yourself thoroughly with the Table of Contents of each one of the volumes listed, and if possible, memorize them. If you have difficulty committing all of the contents to memory you should at least know the contents of the most important reference works. For example, you must remember that Pamphlets 90A, 90B, 91, 211, and 214 of the NFPA are all to be found in *Volume No. NFC-4*, and that "natural ventilation" will be found in the *ASHRAE Handbook of Fundamentals* while "industrial ventilation" will be found in the *ASHRAE Guide Volume Systems*. Study the contents of all of these books carefully. When you are thoroughly familiar with the contents, turn to the index and check the index against the text; this will give you an in-depth feel for each book.

Now, compile a "master table of contents" covering all of the material involved. This "master list of table of contents" will act as a subject guide index and could save the day for you. We have seen candidates taking an open book exam, flipping through page after page, in book after book, looking for the clue to a problem; these candidates always flunk.

Usually your weakness will show in a particular area. You may score high in psychrometrics and do poorly on control wiring. Or you may do very well on questions of a theoretical nature and show a weakness in building code regulations. This will be your clue on where to concentrate. No amount of memorization will help you...know your weakness and study hard in that area!

Whether an exam is "open book" or "closed book" an examination may be phrased in any of a number of styles. Often, the material for an exam, may be gathered by several different persons and comes from many sources. This will be reflected in the style of the questions; sometimes different styles are reflected in one examination owing to the different persons who were involved in writing the exam. Where an editor has been assigned to the job, the exam will be well organized and have an easy style.

Usually, questions will be of the *objective* type. That means you will not be required to do any design drawing or lengthy essay writing. Such questions are true and false, multiple choice, fill-in, etc. In other instances the exam--or portions of it--may require design work (usually of a limited nature), chart plotting, and/or lengthy essay type answers. The present trend is towards objective type examinations answered on machine-scored cards. Most contractor, engineer, journeyman, and civil-service examinations around the country, are using the objective type examination with machine-scoring. In a few cases the largest part of the exam will be objective type supplemented with a small essay type section.

WHAT TO STUDY

Whether the examination is "open book" or "closed book", you must know what to study from. In Figure 1.2 you see a copy of *Notice & Instructions to Examinees*, which was distributed to candidates for a Mechanical Contractor's exam in 1967. In this case the examinee is told that there will be some questions with direct reference to the *Carrier Design Manual*, Part 1, Load Estimating. In other cases, and this is frequently done, you will be given a list of *Recommended Reference*. The recommended reference is bibliography; it tells you what, in the opinion of the examiners, you should study. For an open book exam, you may be given a list of reference material that will be permitted inside of the examination room.

Usually, examination announcements will offer the candidate a list of recommended reference material. Such lists should receive your most careful attention and all books, pamphlets, codes and texts should be procured at the earliest possible time. Often, however, these lists are incomplete, incorrect, or obsolete. There is a tendency to reprint the same list year after year although some of the material has gone out of print or has been superseded by revised editions or new material.

Standard handbooks and reference manuals are usually updated periodically; in an era of fast-breaking new technologies, such new editions will completely obsolete the preceding ones. When a new standard is published, such as the *ASHRAE Guide and Data Book*, or the *National Electric Code*, only the new current edition is obtainable. The previous editions--having become obsolete--go out of print and are removed from available sources.

Writers of examinations as well as instructors for prep courses often neglect to update their material and are inclined to reuse old sources as a matter of convenience. National code revisions are usually adopted into examinations and college or vocational curriculum, long after the old code literature has gone into disuse. It is strongly recommended that examination candidates insist upon detailed instructions for obtaining reference material no longer available from the publisher. When a reference text has been declared out of print by a publisher, it should be removed from the reference list and automatically dropped as a study book for course work.

If a reference text is out of print and cannot be obtained, you should make a point of explaining this to your local examining board as soon as you have discovered it. If you do not receive any satisfaction from the board, you should then go beyond them and bring the matter to the attention of the county manager, or whoever is the higher authority. Sometimes candidates are obligated to take examinations for which no reference material has been available, simply because the exam writers were ignorant of the superseding edition or were using outdated material as a convenience.

TABLE 1.1

LIST OF RECOMMENDED REFERENCES AND STUDY MATERIAL FOR AIR CONDITIONING JOURNEYMAN

1. AIR CONDITIONING TESTING/ADJUSTING/BALANCING, Gladstone, (Engineer's Press)
2. AIR CONDITIONING CONTRACTORS OF AMERICA (ACCA) Pamphlets: Manual B, Manual L, Manual 4, Manual 6, ELECTRICAL COMPONENTS AND SYSTEMS, AUTOMATIC CONTROLS FOR HEATING AND AIR CONDITIONING SYSTEMS.
3. ANSWERS ON BLUE PRINT READING, Graham, (Audels)
4. AUTOPSY ON A COMPRESSOR, Burl C. Brown
5. HOME REFRIGERATION & AIR CONDITIONING, Anderson (Audels)
6. MODERN REFRIGERATION AND AIR CONDITIONING, Althouse, (Goodheart & Wilcox)
7. NATIONAL FIRE CODE (NFPA) Pamphlets: #90A, Air Conditioning, #91, Blower and Exhaust, #31, Oil Burning Equipment, #90B, Warm Air, #96, Commercial Kitchen Exhaust, #54, Fuel Gas
8. PIPEFITTER'S HANDBOOK, Lindsley, (Industrial Press)
9. PLUMBERS AND PIPEFITTERS; MATERIAL, Graham (Audels)
10. PLUMBERS AND PIPEFITTERS; INSTALLATION, Graham (Audels)
11. PRINCIPLES OF AIR CONDITIONING, Lang
12. READING AND INTERPRETING DIAGRAMS IN AIR CONDITIONING AND REFRIGERATION, Mahoney (Reston, Prentice-Hall)
13. SAFETY CODE FOR MECHANICAL REFRIGERATION, ASHRAE/ANSI 15
14. SHEET METAL AND AIR CONDITIONING CONTRACTORS ASSOC.(SMACNA) Pamphlets: HVAC Systems--Duct Construction Standards--Metal and Flexible; Fibrous Glass Duct Construction
15. SHEET METAL WORKERS HANDYBOOK, Graham (Audels)
16. TRANE RECIPROCATING REFRIGERATING MANUAL (The Trane Co.)

NOTE: If you are not able to find a sales outlet for these books, they are all stocked at the Construction Bookstore Inc., West Palm Beach and Gainesville, FL. Call (305) 433-5166 to check prices and availability or to place your order for immediate delivery. In Florida call toll-free, 1-800-432 5540.

Table 1.1 shows a list of Reference and Study Material compiled by the author. This is a basic study list and should be considered minimal for each particular category. If your examination announcement lists other specific books and pamphlets, then, of course, you are responsible for the additional information. Although few manufacturer's pamphlets appear in Table 1.1, (because they are too numerous to list in this limited space), they are often the most important source material.

Manufacturers and trade associations publish pamphlets and books containing excellent, reliable information concerning their products or areas of interest. These publications may cover design, safety, selection, application, installation and maintenance of products ranging from mosaic tile and plaster to boilers, pipe, and pollution control products. You should be thoroughly familiar with the available literature for your particular category. Locally, these may be available at your union hall, supply house, trade group, association, or manufacturer's representative. We suggest that you make every effort to acquire and study as much of this material as possible; it will provide answers to many of the practical questions appearing in your examination. You should also have a good dictionary of the English language.

In addition to listing obsolete references, the official announcement of recommended references may show incomplete or incorrectly spelled titles; or the publisher's or author's name may be misspelled or deleted. When you receive your official list from your municipality, check it against Table 1.1 for hints to correct titles.

CODES, STANDARDS AND STUDY MATERIAL

Bookstores do not generally carry the literature you are required to study from. If your local bookstore does carry some of the the listed titles, good; if they do not carry them in stock, they may be willing to special order them for you. If your study material is not readily available through a local source, you will have to order by mail. In most cases your order will have to be accompanied by a check or money order; sometimes handling and postage charges are also required. In Table 1.2 you will find a list of the names and addresses of the most important associations and publishers of national codes and standards, as well as some of the publishers of other recommended reference texts and handbooks.

Boilers and Pressure Vessel Code: ASME

The ASME Boiler and Pressure Vessel Code is the work of the American Society of Mechanical Engineers; Boiler and Pressure Vessel Committee. This code is independent of ANSI and consequently is not referenced by the ANSI numbering system.

It consists of the following Sections, which are shown with ASME's list price--for the convenience of contractor examination candidates:

Section	Title	Price
I	Power Boiler	$130.00
II	Material Specifications	
	Part A Ferrous	$225.00
	Part B Nonferrous	$220.00
	Part C Welding Rods	$125.00
IV	Heating Boilers	$125.00
V	Nondestructive Examination	$125.00
VI	Care and Operation of Heating Boilers	$65.00
VII	Care of Power Boilers	$85.00
VIII	Pressure Vessels	
	Division 1	$255.00
	Division 2: Alternative Rules	$255.00
IX	Welding Qualifications	$125.00
X	Fiberglass Reinforced Plastic Pressure Vessels	$105.00

Candidates for all contractor categories, and particularly the mechanical categories, should carefully examine the titles and contents of the "Standards" in the recommended reference lists before embarking on a study program. If any of these appear on the reference list circulated by the examining board, you should not hestitate to incorporate them into your study program. Table 1.3 cross index of the standards applying to the mechanical section of the South Florida Building Code; it is typical of the standards adopted by most of the codes around the country.

SFBC & SSBC Versus National Standards

Occasionally a conflict will be found betwen a Municipal Code and a national Standard. When this occurs remember that your local code takes precedence over any other Standard, even a self-adopted one. For example, the SFBC, 4802 (c), calls for a "complete change of air every three minutes; for dry-cleaning plants. The NFPA 32 calls for an "air change every five minutes. The SFBC supersedes the NFPA in South Fla.

Code for Pressure Piping

The Code for Pressure Piping is sponsored by the American Society of Mechanical Engineers and consists of the following major standards:

Section	Title	Price
B31.1	Power Piping	$105.00
B31.2	Fuel Gas Piping	7.50
B31.3	Chemical Plant and Petroleum Refinery Piping	$125.00
B31.4	Liquid Transportation Systems for Hydrocarbons, Liquid Petroleum Gas, Anhydrous Ammonia and Alcohols	$60.00
B31.5	Refrigeration Piping	$50.00
B31.8	Gas Transmission and Distribution	$70.00
B31.9	Building Service Piping	$12.00
B31.11	Slurry Transportation Piping Systems	$70.00

WHAT TO BRING TO THE EXAM ROOM

In most cases the official Instructions to Examinees will list the items you may bring with you, or "must furnish", in addition to the standard reference books. If you require reading glasses, bring two pair ...just in case. Also bring a few colored Fine-Point pens as well as a yellow Highlight marker; these often prove invaluable in the exam room. A pocket scale-rule and two good electronic calculators with extra batteries and a small battery operated desk clock should complete your list.

Table 1.1 will help you select the basic reference texts. In addition you may have some manufacturer's catalog from which you have been accustomed to work, these will be helpful because you are familiar with them. The key to any open book exam is knowing where to find the information: familiarize yourself thoroughly with all recommended reference works--know your code books. If you must memorize anything, memorize the Table of Contents and Index of your most important reference book!

THE DAY OF THE EXAMINATION

Almost as important as the examination day, is the day before. Try to schedule an easy, relaxed day before the exam. Do not do any unnecessary work on exam night; relax and get a good night's sleep. Don't try to cram a lot of studying in on the last night; if you do not know your stuff now, it's too late.

Allow yourself plenty of time to arrive at the examination room a half-hour ahead of start time. This is important advice. Hazardous driving, hurrying, tension of any kind...will affect your thinking and recall faculties. Arrive early--keep confident--keep relaxed.

(Continued on page 30)

TABLE 1.2

NAMES AND ADDRESSES OF IMPORTANT AGENCIES, ASSOCIATIONS AND PUBLISHERS

Manufacturers and trade associations publish pamphlets and books containing excellent, reliable information concerning their products on areas of interest. These publications may cover design, safety, selection, application, installation and maintenance of products ranging from mosaic tile and plaster to boilers, pipe, and pollution control products. You should be thoroughly familiar with the available literature for your particular category. Locally, these may be available at your union hall, supply house, trade group, association, or manufacturer's representative. We suggest that you make every effort to acquire and study as much of this material as possible; it will provide answers to many of the practical questions appearing in your examination.

American Conference of Governmental Industrial Hygienists
Kemper Woods Center
1330 Kemper Meadow Drive Suite 600
Cincinnati, OH 45240

Air Conditioning Contractors of America (ACCA)
1712 New Hampshire Avenue, N.W.
Washington, DC 20009

Air Conditioning and Refrigeration Institute (ARI)
4301 N. Fairfax Drive Suite 425
Arlington, VA 22203

American Gas Association (AGA)
1515 Wilson Blvd
Arlington, VA 22209

American National Standards Institute Inc. (ANSI)
(ANSI: Formerly American Standards Association,
ASA and United States of American Standards Institute)
11 W. 42nd Street 13th Floor
New York, NY 10036

American Society of Heating, Refrigeration & Air-Conditioning Engineers (ASHRAE)
1791 Tullie Circle N.E.
Atlanta, GA 30329

American Society of Mechanical Engineers (ASME)
345 East 47th Street
New York, NY

American Water Works Association (AWWA)
6666 W. Quincy Avenue
Denver, CO 80235

American Welding Society (AWS)
550 N.W. LeJeune Road
Miami, Fl 33135

Building Officials and Code Administrators (BOCA)
4051 W. Flossmoor Road
Country Club Hills, IL 60478

Carrier Air Conditioning
Carrier Parkway
P.O. Box 4808
Syracuse, NY 13221

International Conference of Building Officials
5360 S. Workman Mill Road
Whittier, CA 90601

McGraw-Hill Book Company
11 West 19th Street
New York, NY 10011-4285

National Bureau of Standards, Products Standard Section
Center for Building Technology
Washington, DC 20234

National Environmental Balancing Bureau
8224 Old Courthouse Road
Vienna, VA 22180

National Fire Protection Association (NFPA)
1 Batterymarch Park
Quincy, MA 02269

Sheet Metal & Air Conditioning Contractor's National Association (SMACNA
4201 Lafayette Center Drive
Chantilly, VA 22021

Southern Building Code Congress
900 Montclair Road
Birmingham, AL 35213

Superintendent of Documents
U.S. Government Printing Office (GPO)
Washington, DC 20402

The Trane Company
LaCrosse, WI 54601

Underwriter's Laboratories, Inc. (UL)
333 Pfingsten Road
Northbrook, Il 60062

ACCA MANUALS

Description and prices (1987) of important manuals of the Air Conditioning Contractors of America

---

Manual B

Principles of Air Conditioning
Copyright 1970 (First Edition)

Describes the fundamental principles of heating and cooling in plain, non-technical language. Illustrates how heat travels and how air flows through ducts. Contains a discussion on heat loss and heat gain in buildings and a complete glossary.

Member price $5.00 Non-member price $10.00

Manual C

What Makes a Good Air Conditioning System?
Copyright 1966 (First Edition)

Describes and illustrates essential elements of heating and air conditioning equipment. Defines the type of electric motors used to drive fans and blowers. Explains how to select a heating unit which will have the proper air handling characteristics essential to good year-round air conditioning systems.

Member price $5.00 Non-member price $10.00

Manual D

Duct Design for Residential Winter and Summer Air Conditioning and Equipment Selection
Copyright 1984 (Second Edition)

Discusses factors affecting selection of equipment and systems design. Describes:

- economics of duct design,
- duct system design principles and procedures,
- duct systems and heating and cooling equipment,
- extended plenum, radial and flexible duct design examples,
- glossary of terms and relevant tables and charts.

Member price $8.00 Non-member price $16.00

Manual E

Room Air Distribution Considerations
Copyright 1965 (First Edition)

Defines the essentials of proper room air distribution, recommends ways to determine the best air distribution patterns, and provides practical procedures for selecting and locating outlets and returns.

Member price $5.00 Non-member price $10.00

Manual L

Recommended Installation Practices and Recommende Municipal Code
Copyright 1969

Describes basic design, system selection, and installatio practices which are applicable to all types of structures. Pre sents a recommended warm air heating and cooling code fc municipalities. Provides representative questions from reg istration examinations for contractors and mechanics.

Member price $5.00 Non-member price $10.0

Manual N

Load Calculation for Commercial Summer and Winter Ai Conditioning (Using Unitary Equipment)
Copyright 1983 (Second Edition)

Provides simplified, practical procedures for calculating bot heat loss and heat gain in commercial structures using unitar air conditioning equipment and systems. Describes:

- inside and outside design conditions,
- solar radiation heat through glass,
- infiltration and ventilation,
- dehumidification and humidification,
- load calculation examples.

Member price $8.00 Non-member price $16.0

Manual P

Psychrometrics: Theory and Applications
Copyright 1985 (First Edition)

A practical, how-to manual that explains psychrometric pro cesses and calculations, and how to use the psychrometr chart (10 four-color psychrometric charts are included wit each manual). Covers:

- Room (or zone) design dry bulb temperature and relativ humidity and how it will be maintained by the coolin equipment,
- the conditions needed to maintain the indoor design dr bulb temperature and relative humidity,
- how the psychrometric chart can be used to analyze th particular processes associated with the various types heating and cooling systems.

Member price $8.00 Non-member price $16.

Manual J
Load Calculation for Residential Winter and Summer Air Conditioning
Copyright 1986 (Seventh Edition)

Provides a simple, accurate procedure for estimating the heat loss and heat gain for conventional residential structures. Explores the underlying assumptions and physical principles of the calculation procedures. Provides example problems and information on:

- mobile home load calculations,
- multi-zone systems,
- multi-family structures,
- energy consumption,
- infiltration,
- operating costs.

Member price $8.00 Non-member price $16.00

Manual H
Heat Pump Systems: Principles & Applications
Copyright 1984 (Second Edition)

Includes a complete discussion of:

- heat pump selection, performance and operation,
- basic principles such as integrated capacity, COP, theoretical and operating balance points and sizing considerations,
- types of heat pump systems,
- design procedures for both residential and commercial applications,
- installation practices including equipment location, accessibility, refrigerant piping, controls and power wiring.

Member price $8.00 Non-member price $16.00

Manual G
Selection of Distribution Systems
Copyright 1963 (First Edition)

Lists factors which must be considered before proper system selection is made. Describes types of duct systems and how they can be used in residences and apartments. Illustrates performance characteristics of various systems. Describes the effects of climate on system selection. Explains the achievement of system balance through zoning and provides an illustrated guide for system selection.

Member price $5.00 Non-member price $10.00

Manual 4
Installation Techniques for Perimeter Heating and Cooling
Copyright 1964 (Tenth Edition)

Provides recommendations on installation and construction practices for perimeter heating and cooling systems in slab floor and crawl space structures.

Member price $5.00 Non-member price $10.00

Manual Q
Equipment Selection and System Design Procedures for Commercial Summer and Winter Air Conditioning (Using Unitary Equipment)
Copyright 1977 (First Edition)

Discusses factors affecting system design, types of systems, equipment selection, refrigerant piping, duct system design, and useful formulas. Also includes a glossary and a metric conversion table.

Member price $8.00 Non-member price $16.00

Introduction to the Installation of Residential Ducted Heating and Air Conditioning Systems
Copyright 1972

A manual for the beginning craftsman. Describes equipment, systems, ducts and fittings, piping and fittings, tools and installation procedures. Contains comprehensive drawings and sketches.

Member price $8.00 Non-member price $16.00

Automatic Controls for Heating and Air Conditioning Systems
Copyright 1966 (First Edition)

Provides basic information necessary to understand, sell, install, and service automatic controls for heating and air conditioning systems. Describes construction, function, and operation of automatic control systems.

Member price $8.00 Non-member price $16.00

Residential Heating and Air Conditioning Systems—Minimum Installation Standards
Copyright 1973 (Second Edition)

Provides minimum installation standards based on accepted practices and authoritative research data. Covers system design, equipment ducts, piping, chimneys, vents, filters, humidification sound, wiring, start-up, adjustment, and instructions to owner. (Published jointly by ACCA and SMACNA.)

Member price $3.00 Non-member price $6.00

Manual 6
Adjusting Air Conditioning Systems for Maximum Comfort
Copyright 1979 (Third Edition)

Describes how winter, summer, or year-round air conditioning systems are adjusted for comfort air circulation. Gives step-by-step procedures for adjusting thermostats and controls, and for system balancing. Offers recommendations to contractors for providing annual adjustment service to customers.

Member price $5.00 Non-member price $10.00

SMACNA MANUALS

Description and prices (1987) of important manuals of the Sheet Metal and Air Conditioning Contractor's National Association Inc.

The SMACNA Standards are published by the Sheet Metal and Air Conditioning Contractor's National Association Inc. The SMACNA standards or "pamphlets" as they are often called, contain vital information and are frequently used as source material by exam writers. The most commonly used for the HAC&V industry are:

HVAC Systems—Duct Design
Revised design manual for commercial and light industrial HVAC systems. Offers options—methods, materials and construction—for solving energy costs and conservation.
1981/256 pp/**Code:** SMA33
List: $58.00 **Member:** $34.80

HVAC Systems—Duct Design Tables and Charts
Breakout of Chapter VI of the SMACNA HVAC Duct System Manual. Compact collection of data for ducts and fittings in both inch-pound and SI units of measurement.
1981/80 pp/**Code:** SMA34
List: $23.00 **Member:** $13.80

HVAC Duct Construction Standards—Metal and Flexible
Primarily used for commercial and institutional projects, but applicable for residential and certain industrial work. Prescribes construction detail alternatives for uncoated and galvanized steel, aluminum and stainless steel ductwork consisting of various sections, transitions, elbows, fittings, and accessories.
1985/220 pp/**Code:** SMA32
List: $58.00 **Member:** $34.80

HVAC Air Duct Leakage Test Manual
Companion to HVAC Duct Construction Standards; contains duct construction leakage rates for sealed and unsealed ducts, test procedures, types of test apparatus, test set-up and sample leakage analysis.
1985/44 pp/**Code:** SMA29
List: $29.00 **Member:** $17.40

Energy Conservation Guidelines
Familiarizes HVAC contractors with the potential energy savings available in new and existing buildings.
1984/151 pp/**Code:** SMA24
List: $23.00 **Member:** $13.80

Fibrous Glass Duct Construction Standards
Covers pressure sensitive tape standards, fibrous glass board, duct and fitting fabrication, seam and joint closures, reinforcements with tee bars, channels and tie-rods, and hangers and supports.
1979/124 pp/**Code:** SMA26
List: $35.00 **Member:** $21.00

Round Industrial Duct Construction Standards
A standardized, engineered basis for design and construction of industrial ducts. Comprehensive details for duct construction and support systems.
1977/260 pp/**Code:** SMA43
List: $58.00 **Member:** $34.80

Thermoplastic Duct (PVC) Construction Manual
Construction details, a model specification for preparing project documents for PVC fume exhaust systems and general information on plastic materials.
1974/106 pp/**Code:** SMA44
List: $23.00 **Member:** $13.80

HVAC Systems—Testing, Adjusting and Balancing
The definitive manual on air and hydronic balancing and adjusting. Allows contractors to supervise the balancing of any system.
1983/256 pp/**Code:** SMA36
List: $58.00 **Member:** $34.80

Retrofit of Building Energy Systems and Processes
A guide to total building retrofit and systems with emphasis on energy conservation.
1982/336 pp/**Code:** SMA42
List: $63.00 **Member:** $37.80

ASHRAE MANUALS

Description and prices (1987) of important manuals of the Society of Heating Refrigerating & Air Conditioning Engineers.

ASHRAE Handbook of Fundamentals	$100.00
ASHRAE Handbook of Applications	$100.00
ASHRAE Handbook of Equipment	$100.00
ASHRAE Handbook of Systems	$100.00
ASHRAE Handbook of Refrigeration	$100.00
Standard 15, Safety Code for Mechanical Refrigeration	$30.00

Many other ASHRAE publications may be found in the current ASHRAE Publications Catalog.

NFPA FIRE CODES

Description and prices (1987) of important code books of the National Fire Protection Association.

Listed here are those pamphlets most frequently referred to in contractor and journeyman examinations. The NFPA committees update these pamphlets periodically. The examination candidates should make certain that they use the code date listed in the official examination instructions.

NFPA 12: *Carbon Dioxide Extinguishing Systems*
Covers design, installation, testing, approval, operation, and maintenance for total flooding, local application, and hose line systems. 1985
Item No. B7-NFPA-12 $11.50 (Members: $10.35)

NFPA 12A: *Halon 1301 Fire Extinguishing Systems*
Provides minimum requirements for design, installation, testing, inpsection, and maintenance. 1985
Item No. B7-NFPA-12A $13.50 (Members: $12.15)

NFPA 12B: *Halon 1211 Fire Extinguishing Systems*
Provides the minimum requirements for design, installation, testing, inspection, and maintenance. 1985
Item No. B7-NFPA-12B $11.50 (Members: $10.35)

NFPA 12C-T: *Halon 2402 Fire Extinguishing Systems*
Tentative standard contains minimum requirements for use in confined and unoccupied outdoor areas that do not pose hazards to personnel. 1983
Item No. B7-NFPA-12C-T $10.50 (Members: $9.45)

NFPA 13: *Installation of Sprinkler Systems*
Standard for the installation and testing procedures for fire protection sprinkler systems, including wet pipe, dry pipe, deluge, preaction, and more. 1985
Item No. B7-NFPA-13 $12.50 (Members: $11.25)

NFPA 13A: *Inspection, Testing and Maintenance of Sprinkler Systems*
Includes practical advice for managers and property owners on how to keep sprinkler systems in proper operating condition. 1981
Item No. B7-NFPA-13A $10.50 (Members: $9.45)

NFPA 13D: *Installation of Sprinkler Systems in One- and Two-Family Dwellings and Mobile Homes*
Covers the design and installation of automatic sprinkler systems and information on water supply, system design, and other important technical considerations. 1984
Item No. B7-NFPA-13D $9.50 (Members: $8.55)

NFPA 30: *Flammable and Combustible Liquids Code*
Requirements for tank storage, piping, valves and fittings, container storage, industrial plants, bulk plants, and processing plants. 1984
Item No. B7-NFPA-30 $11.50 (Members: $10.35)

NFPA 30A: *Automotive and Marine Service Station Code*
General provisions for piping, fuel dispensing systems, stations inside buildings, and operational requirements. 1984
Item No. B7-NFPA-30A $9.50 (Members: $8.55)

NFPA 17: *Dry Chemical Extinguishing Systems*
Minimum requirements plus discussion of total flooding, local application, hand hose line, and pre-engineered systems. 1985
Item No. B7-NFPA-17 $10.50 (Members: $9.45)

NFPA 20: *Installation of Centrifugal Fire Pumps*
Provides guidelines for design, installation, and maintenance of centrifugal fire pumps, pump drivers, horizontal pumps, and vertical shaft turbine-type pumps. 1983
Item No. B7-NFPA-20 $11.50 (Members: $10.35)

NFPA (Continued)

NFPA 21: *Steam Fire Pumps*
Applies to the National Standard steam fire pumps. Includes information about pump operation, steam supply, testing, maintenance, and more. 1982
Item No. B7-NFPA-21 $9.50
(Members: $8.55)

NFPA 22: *Water Tanks for Private Fire Protection*
Requirements for gravity and suction tanks, towers, foundations, piping, valves, heating, pressure tanks, and tank insulation. 1984
Item No. B7-NFPA-22 $11.50
(Members: $10.35)

NFPA 24: *Installation of Private Fire Service Mains*
Gives details on yard piping that supplies water to automatic sprinkler systems, yard hydrants, standpipes, and other systems. 1984
Item No. B7-NFPA-24 $10.50
(Members: $9.45)

NFPA 26: *Supervision of Valves Controlling Water Supplies for Fire Protection*
Outlines the recommendations for identification and supervision of valves, including valve seals and tags, and valve index boards. 1983
Item No. B7-NFPA-26 $9.50
(Members: $8.55)

NFPA 27: *Private Fire Brigades*
Gives recommendations for organizing, managing, training, inspection, maintenance, and equipping private fire brigades. 1981
Item No. B7-NFPA-27 $9.50
(Members: $8.55)

NFPA 31: *Installation of Oil Burning Equipment*
Requirements for stationary and portable oil burning equipment, tanks, piping, and accessories. 1983
Item No. B7-NFPA-31 $10.50
(Members: $9.45)

NFPA 32: *Drycleaning Plants*
Outlines specific safeguards to be applied to drycleaning and dry dyeing operations. 1985
Item No. B7-NFPA-32 $9.50
(Members: $8.55)

NFPA 50: *Bulk Oxygen Systems at Consumer Sites*
Provides recommendations for location, distance between bulk systems and exposures, containers, and associated equipment. 1985
Item No. B7-NFPA-50 $9.50
(Members: $8.55)

NFPA 50A: *Gaseous Hydrogen Systems at Consumer Sites*
Requirements for containers, safety relief devices, piping, and other components. 1984
Item No. B7-NFPA-50A $9.50
(Members: $8.55)

NFPA 50B: *Liquefied Hydrogen Systems at Consumer Sites*
Requirements for containers,supports, marking, safety releases, piping and other components. 1985
Item No. B7-NFPA-50B $9.50
(Members: $8.55)

NFPA 51: *Oxygen-Fuel Gas Systems for Welding, Cutting, and Allied Processes*
Applies to acetylene and oxygen cylinder storage and use. Covers MAP, other stable gases, and acetylene generation. 1983
Item No. B7-NFPA-51 $9.50
(Members: $8.55)

NFPA 51A: *Acetylene Cylinder Charging Plants*
Provides safety requirements for the design, construction and installation of acetylene cylinder charging plants. 1984
Item No. B7-NFPA-51A $9.50
(Members: $8.55)

NFPA 34: *Dipping and Coating Processes Using Flammable or Combustible Liquids*
Requirements for coating, finishing, cleaning, and treating processes which use flammable or combustible liquids. 1982
Item No. B7-NFPA-34 $10.50
(Members: $9.45)

NFPA 51B: *Cutting and Welding Processes*
Practices and precautions for cutting and welding processes involving the use of electric arcs and oxy-fuel gas flames. 1984
Item No. B7-NFPA-51B $9.50
(Members: $8.55)

NFPA 52: *Compressed Natural Gas (CNG) Vehicular Fuel Systems*
Applies to general CNG and equipment qualification, engine fuel systems, storage and dispensing systems. 1984
Item No. B7-NFPA-52 $10.50
(Members: $9.45)

NFPA 53M: *Fire Hazards in Oxygen-Enriched Atmospheres*
Covers materials, system designs, and fire control. This manual also documents actual reports of fire experience in OEAs. 1985
Item No. B7-NFPA-53M $11.50
(Members: $10.35)

NFPA 54: *National Fuel Gas Code*
Requirements for the safe design, installation, operation, and maintenance of gas piping in buildings and gas appliances in residential, commercial, and industrial applications. 1984
Item No. B7-NFPA-54 $10.50
(Members: $9.45)

NFPA 56F: *Nonflammable Medical Gas Systems*
Safety practices for piped oxygen, nitrous oxide, compressed air, and other nonflammable medical gases in hospitals and other medical facilities. 1983
Item No. B7-NFPA-56F $10.50
(Members: $9.45)

NFPA 59: *Storage and Handling of Liquefied Petroleum Gases at Utility Gas Plants*
Provides for the safe design, construction, and operation of LP-Gas equipment at plants supplying LP-Gas/air mixtures for utility application. 1984
Item No. B7-NFPA-59 $10.50
(Members: $9.45)

NFPA 85A: *Prevention of Furnace Explosions in Fuel-Oil and Natural Gas-Fired Single Burner Boiler-Furnaces*
Applies to boilers with fuel input greater than 12,500,000 Btuh (3663 KW) that use single burners firing natural gas, fuel oil, or both. Includes information on design, installation, operation, and maintenance. 1982
Item No. B7-NFPA-85A $10.50
(Members: $9.45)

NFPA 85B: *Prevention of Furnace Explosions in Natural Gas-Fired Multiple Burner Boiler-Furnaces*
Requirements for the design, installation, operation, and maintenance of specified boiler-furnaces, their fuel-burning systems, and related control equipment. 1984
Item No. B7-NFPA-85B $10.50
(Members: $9.45)

NFPA 85D: *Prevention of Furnace Explosions in Fuel Oil-Fired Multiple Burner Boiler-Furnaces*
Outlines equipment requirements, sequencing of operations, and interlock and alarm requirements for specified boiler-furnaces. 1984
Item No. B7-NFPA-85D $10.50
(Members: $9.45)

NFPA 85E: *Prevention of Furnace Explosions in Pulverized Coal-Fired Multiple Burner Boiler-Furnaces*
Requirements for the design, installation, and operation of pulverized coal-fired multiple burner boiler-furnaces, fuel-burning systems, and control equipment. 1985
Item No. B7-NFPA-85E $10.50
(Members: $9.45)

NFPA 85F: *Installation and Operation of Pulverized Fuel Systems*
Requirements for the design, installation, operation, maintenance, and personnel safety around pulverized fuel systems.
Item No. B7-NFPA-85F $10.50
(Members: $9.45)

NFPA 85G: *Prevention of Furnace Implosions in Multiple Burner Boiler-Furnaces*
Applies to the design, installation, and operation of boiler-furnaces to prevent furnace implosions and to promote operating safety. 1982
Item No. B7-NFPA-85G $9.50
(Members: $8.55)

NFPA 86: *Ovens and Furnaces*
Applies to "Class A/B" ovens or furnaces—heat utilization equipment operating at atmospheric pressures and used by industry for the processing of materials. Provides requirements for location, construction, operation, heating system, ventilation, safety controls, and fire protection. 1985
Item No. B7-NFPA-86 $11.50
(Members: $10.35)

NFPA 86C: *Industrial Furnaces Using a Special Processing Atmosphere*
Covers "Class C" industrial furnaces which utilize a special processing atmosphere, including salt bath and integral quench furnaces. Provides requirements for location, construction, heating system, safety controls, operation, and fire protection. 1984
Item No. B7-NFPA-86C $11.50
(Members: $10.35)

NFPA 90A: *Installation of Air Conditioning and Ventilating Systems*
Installation requirements for air conditioning and ventilating systems that restrict the spread of smoke, heat, and fire through duct systems, minimize ignition sources, and make the systems usable for emergency smoke control. 1985
Item No. B7-NFPA-90A $10.50
(Members: $9.45)

NFPA 90B: *Installation of Warm Air Heating and Air Conditioning Systems*
Provides installation requirements for supply ducts, controls, clearances, heating panels, return ducts, air filters, heat pumps, and other components. Applies to one- and two-family dwellings or spaces not exceeding volumes of 25,000 cubic feet. 1984
Item No. B7-NFPA-90B $9.50
(Members: $8.55)

NFPA 91: *Blower and Exhaust Systems for Dust, Stock and Vapor Removal or Conveying*
Requirements for fans, ducts, duct clearances, design, and dust collecting systems for removal or conveying of flammable vapors, corrosive fumes, dust, stock, and refuse. 1983
Item No. B7-NFPA-91 $9.50
(Members: $8.55)

NFPA 96: *Removal of Smoke and Grease-Laden Vapors from Commercial Cooking Equipment*
Covers the design, installation, and use of exhaust system hoods, grease removal devices, ducts, dampers, air moving devices, and fire extinguishing equipment. 1984
Item No. B7-NFPA-96 $9.50
(Members: $8.55)

NFPA 97M: *Standard Glossary of Terms Relating to Chimneys, Vents and Heat-Producing Appliances*
Defines terms with the purpose of achieving uniformity for standards involving chimneys, gas vents, and heat-producing appliances. 1984
Item No. B7-NFPA-97M $9.50
(Members: $8.55)

TABLE 1.3

CROSS INDEX OF THE MAJOR STANDARDS PERTAINING TO THE MECHANICAL SECTIONS OF THE SOUTH FLORIDA BUILDING CODE (1984)

Column 1 gives the adopted standard, column 2 gives the construction covered, Column 3 gives the paragraph where the SFBC reference appears.

STANDARD	CONSTRUCTION COVERED	SFBC SECTION
ANSI A 17.1	Elevators and escalators	3201.2
NFPA 90 A	Fire resistance protection	3703.6
NFPA 13	Sprinkler systems	3801.2
NFPA 14	Standpipes	3803.1
NFPA 22	Pressure tanks	3804.4
NFPA 54	Vent connections for gas appliance	3905.2
NFPA 90 B	Smoke Pipe clearances	3905.3
NFPA 89 M	Clearances for heat producing appliances	4001.4
NFPA 31	Oil burning equipment	4004
ASME	"Boiler & Pressure Vessel Code" boilers	4006
NFPA 30	Flammable and combustible liquids	4102
NFPA 31	Oil burning equipment	4102
NFPA 90 A	Nonresidential air conditioning	4103
NFPA 90 B	Residential warm air and air conditioning	4103
NFPA 96	Ventilation & restaurant cooking equipment	4103
NFPA 91	Blowers and exhausts	4103
NFPA 664	Woodworking plant dust exhaust	4103.5
NFPA 33	Spray finishing: paint booths	4107.2
NFPA 34	Dip tanks	4107.3

TABLE 1.3 (Continued)

STANDARD	CONSTRUCTION COVERED	SFBC SECTION
C 272	PVC Schedule 40 condensate drains	4606.7
Fla. Health Dept.	Closed well systems	4616.4
NFPA 54	Installation of gas appliances & piping	4702.1
NFPA 58	Storage of LP gas	4702.2
NFPA 33 NFPA 90 A NFPA 90 B NFPA 91 NFPA 96 ASHRAE Guides	Forced ventilation: outside air "Standards of good practice"	4801.2
ANSI B 9.1 ANSI B 31.1 ANSI B 31.2 NFPA 90 B NFPA 214 NFPA 90 A	Air conditioning and refrigeration	4902
ASHRAE Guides	"Standards of good practice	4902

HOW TO TAB THE REFERENCE BOOKS

The importance of index tabbing of the reference materials used in the exam room cannot be over stressed. Every second gained during the exam is a valuable step towards passing. The index system used here has been carefully designed to provide the utmost workability under extreme time pressure. All tables of contents and indexes are red tabs, general materials are clear tabs, and special reference tables and charts are yellow color, for immediated identification.

1. From your local office supply store, purchase the required number of ALCO SURE STICK INDEX TABBING, size 1/3", stock numbeber 10381. Order one box of red and yellow, and as many boxes of clear as needed. One box equals approximately 75 tabs

2. Type or letter the index label as per instructions for individual books. Use three lines of label for each legend as in- indicated below, and

   HEATING CONTROLS INDEX
   HEATING CONTROLS INDEX ← Tear here

   COOLING SYSTEMS
   COOLING SYSTEMS ← Tear here

   type each legend twice, directly beneath one another.

3. Tear off strip, allowing three lines, and fold the blank line inside, allowing the typed legend to appear on both sides as as shown below.

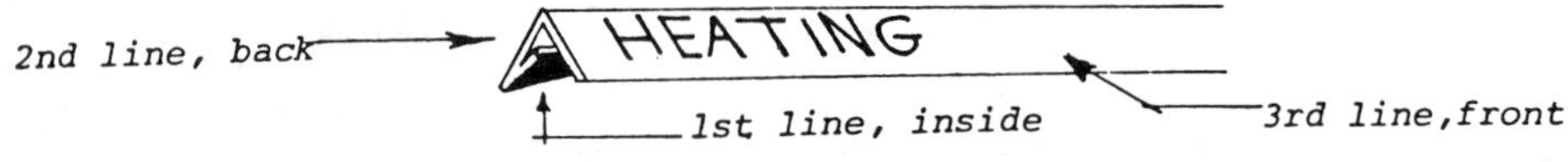

   This three thickness fold is necessary to keep the index label from slipping out of the Tab.

4. Insert folded label into Tab and cut Tab to fit label.

5. Review book tabbing instructions for each book and make a pencil mark on each page (right-hand) to receive a Tab Index, keeping a distance of about 3/8" between the vertical dimensions of each Tab.

6. Line up the Tab on the exact place on the page-making sure you are holding only one page--bend the skirt back sharply to remove backing paper from inside. Press skirt to sheet firmly.

7. The importance of highlighting and tabbing your books cannot be over emphasized. Use a yellow or blue highlight marker and follow the instructions given. This will give you ready-access to important parts of this book.

The following instructions are for tabbing the Trane Air Conditioning Manual 1965, Trane Reciprocating Refrigeration Manual 1971, Modern Air Conditioning and Refrigeration 1982, the South Florida Building Code 1984 and Broward edition. Also included are instructions for highlighting the Safety Code for Mechanical Refrigeration ASHRAE 15-78, and the South Florida Building Code. Use these as a model for highlighting and tabbing your local or regional code books and other required reference books.

TRANE AIR CONDITIONING MANUAL 1965	TAB COPY	PAGE NUMBER	TAB COLOR
	Contents	ix	red
	Basic theory	1	clear
	Heat gain	17	clear
	Psychrometrics	55	clear
	Exp. gas, liquid	57	clear
	Ref. theory	129	clear
	Maintenance	193	clear
	Absorption	213	clear
	Towers	245	clear
	Fans	259	clear
	Ducts	282	clear
	Systems	305	clear
	Temp. conversion	357	yellow
	Steam tables	359	yellow
	Coef: k & C	361	yellow
	Coef: U	363	yellow
	People	373	yellow
	Motors	373	yellow
	OA	375	yellow
	Glass	377	yellow
	Equiv. Δt	383	yellow
	Peaking the load	393	yellow
	Properties of air	395	yellow
	Piping	415	yellow
	Ducts	437	yellow
	Index	449	red

TRANE RECIPROCATING REFRIGERATION MANUAL 1977			
	Contents	V	red
	Refrigerant	1	clear
	Evaporator	9	clear
	Compressor	13	clear
	Condenser	29	clear
	Controls	47	clear
	Piping	76	clear
	Hangers	113	clear
	Testing	119	clear
	Service	123	clear
	SI Metric	137	yellow
	Dictionary	165	yellow
	Index	175	red

MODERN REFRIGERATION AND AC	TAB COPY	PAGE NUMBER	TAB COLOR
	Ref fundamentals	7	clear
	Basic Ref systems	73	clear
	AC Fundamentals	641	clear
	Basic AC systems	673	clear
	Heating, Humidifying	699	clear
	Cooling, Dehumidifying	747	clear
	Air distribution	765	clear
	Heat Pumps	801	clear
	Controls, instruments	837	clear
	Solar	933	clear
	Tables, conversions	947	yellow
	Dictionary	971	yellow
	Index	985	red

SOUTH FLORIDA BUILDING CODE, DADE COUNTY EDITION 1984			
	Index	First pink page	red
	Title page	First white page	red
	Administrative	1-1	clear
	Enforcement	2-1	clear
	Chimneys	39-1	clear
	Heating systems	40-1	clear
	Solar	40-5	clear
	Ventilating Ducts	41-4	clear
	Kitchen exhaust	41-5	clear
	Condensate drains	46-44	clear
	Gas	47-1	clear
	Mechanical ventilation	48-1	clear
	Air conditioning	49-1	clear

SOUTH FLORIDA BUILDING CODE, BROWARD COUNTY			
	Index	First pink page	red
	Title page	First white page	red
	Administrative	1-1	clear
	Enforcement	2-2	clear
	Chimneys	39-1	clear
	Heating Systems	40-1	clear
	Solar	40-3	clear
	Condensate drain	46-23	clear
	Gas	47-1	clear
	Mechanical Systems	48-1	clear
	A. Inspections	48-3	clear
	B. Mech ventilation	48-3	clear
	C. AC & reference	48-5	clear
	D. Ductwork	48-5	clear
	E. Fire dampers	48-7	clear
	F. Kitchen exhaust	48-7	clear
	G. Piping	48-9	clear
	H. Insulation	48-13	clear
	Energy Conservation	CHAPTER 49	clear
	A. Definition	2-1	clear
	B. Design conditions	3-1	clear
	C. Systems	4-1	clear

HIGHLIGHTING

SAFETY CODE FOR MECHANICAL REFRIGERATION, ASHRAE STANDARD 15-89

All references are noted by section number, not page number. Use a yellow or pink highlight marker and mark all sections as shown below.

SECTION	DESCRIPTION
4	Caption
4.1	"
4.1.1	"
4.1.2	"
4.1.2.1	"
4.1.2.1	"
4.1.2.2	"
4.1.2.3	"
4.1.2.4	"
4.2	"
4.2.1.1	"
4.2.1.2	"
5	"
Figure 1	Table Heading
Table 1	Table Heading
6.2	Caption
6.2.1	"
6.2.2	"
6.2.3	"
6.2.4	"
6.4	"
Table 2	Table Heading *
Table 3	Table Heading *
7.1.2	¶
7.1.3	"
7.1.4	"
7.2	Caption
Table 4	Table Heading
7.4.1.2.3	¶
7.4.2	Caption
7.4.2.2	¶
9	Caption
9.1.1	¶
9.4	Caption
9.4.5	First 4 lines & formula *
9.4.8	All
9.4.8.1	"
9.4.8.2	First 6 lines
9.4.8.3	First 8 lines
10	Caption
10.5.1	All
10.9.1	"
10.10	"

SECTION	DESCRIPTION
10.11	All
10.12	Caption
10.12.1	All
10.12.2	"
10.12.3	"
10.13	Caption
10.13.2	First Sentence
10.13.2.1	All
10.13.3	"
10.13.4	"
10.13.5	"
10.13.6.1	Caption, Formula *
10.13.6.2	Caption, Formula a), and b) *
10.14 f)	¶
10.14 h)	All
10.14 i)	¶
11	Caption
11.1	All
11.2	Caption and ¶
11.2.1	¶
11 3	"
12.1	"
12.2	"
12.3	"
12.5	"
12.6	"
12.7	"
12.8	"
12.9	"
12.9.1	"
12.10	"
12.11	All
12.12	¶
Appendix B	Caption
B.2	"
B.3	"
B.10	"
B.16	"

HIGHLIGHTING SOUTH FLORIDA BUILDING CODE

1984 EDITION

PAGE NO.	PARAGRAPH
1-1	102 PURPOSE
2-2	201.3 Paragraph
2-3	(2), (f) Paragraphs
13-1	1301.1 "Group H Occupancy", Paragraph
39-1	3901.3 Definitions (complete)
39-2	(b) WALLS AND FLUE LINING, (1) Paragraph
39-3	(c),(d),(e),(f),(g) 3903 TYPE B FLUES OR VENTS
39-4	3905.1 (including table)
40-1	Heading; HEAT PRODUCING APPARATUS
40-2	4006.2 (a),(b),(c),(d),(e) 4006.3 Caption only 4006.4 Caption only
40-3	4006.6 Caption and (a) 4007 Heading; INCINERATORS
40-4	4008 Heading; SOLAR ENERGY SYSTEMS
40-5	Top of page, (8), (e) 4008.2 (a), (b), (c), 3 4008.3 (a)
40-6	Top of page, (1), (b) 4008.4 (b)
41-2	4102.3 (a)
41-3	(c), (f)
41-4	(2), (m), (o), Paragraphs 4103 Heading; VENTILATING DUCTS
41-5	4103.2 (b), (d), (f) 4103.3 Caption, (b) LOCATION (1), (2), (3)
41-6	(d) HOOD DESIGN (1), (2) (e) DUCTS, (3), (4), (6), (7),

PAGE NO.	PARAGRAPH
46-07	4606.7 AIR CONDITIONING CONDENSTATE, Entire section.
48-1	Heading; MECHANICAL VENTILATION 4802 Caption, REQUIREMENTS BASED ON USE and entire section. Make notes
48-2	Entire page. Make notes.
49-1	Heading; AIR CONDITIONING & REFRIGERATION 4901.1 (b) Paragraph 4901.2 (a) Paragraph 4901.3 Entire section.
49-2	4903.1 Paragraph 4903.5 Caption and (a), (b) 4903.6 (a), (b)
49-3	4903.7 Caption; OUTSIDE AIR, and entire section. 4903.10 Caption only

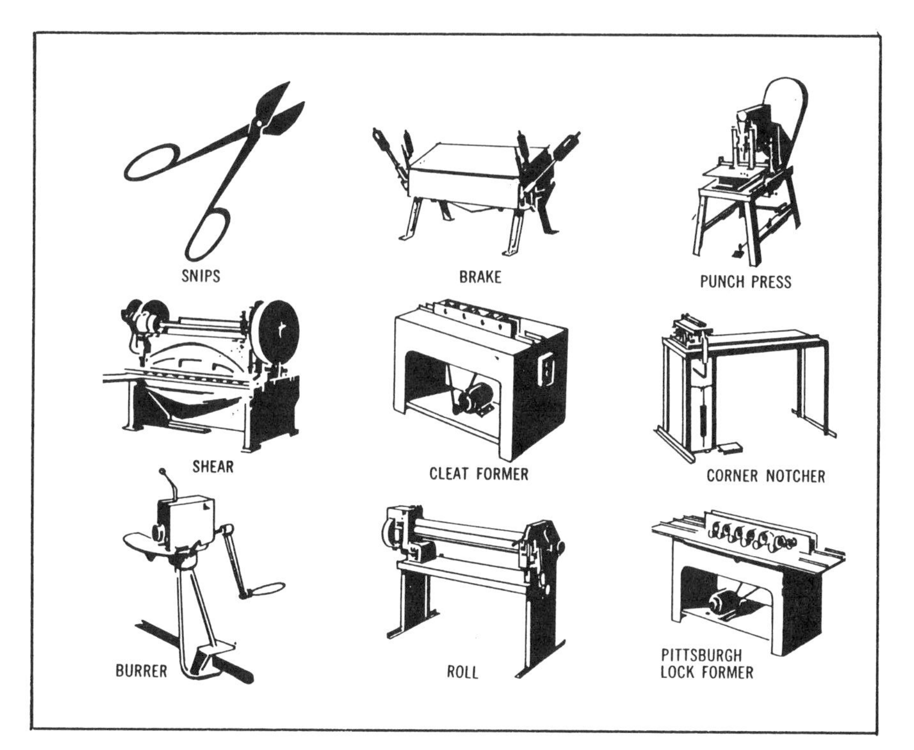

FIGURE 1.1.A

IDENTIFYING SHEET-METAL SHOP EQUIPMENT

Concentrate on the examination. Read the directions carefully, then read them again. If there are verbal instructions, listen attentively. Maintain your sense of organization: get an overview of the exam before you start work and try to judge how much time you can spend on each section.

Keep your work neat. Do not sacrifice a properly marked paper in the interest of speed. Mark each answer carefully and accurately in the proper space provided, especially on machine-scored cards. Mechanical scoring devices have no concept of your good intentions; many candidates have failed examinations because of improperly marked papers.

Examinations are often weighted, i.e., a greater value is placed on one part of the examination than another. If this is the case it will be so stated in your instructions. Try to pace yourself accordingly and spend an equivalent amount of time to the given weight or value. Do not spend half of your allowable time on a section of the exam that is weighted at 25%.

Examinations are seldom graded on a penalty system, therefore you should answer every questions; you are not penalized for an omission. If you do not know an answer, guess at it and take your chances...a chance is better than no answer.

If you find a section of the examination particularly puzzling, move on to the next section, leaving the difficult one to return to later. After you have completed all the easy sections, go back and clean up the tough ones.

Use all of the time allotted to you. If the time limit for the exam is 4 hours, do not leave the room in 3 hours. Spend all of the remaining time reviewing your answers: if you find one error, it was worth the extra hour. But remember, do not plan to do a total reevaluation of your answers. Just review...make certain you marked every question, wrote each answer in the proper place, said what you meant in each case, and clearly understood each question.

If there is a time limit for each section (examinations are sometimes organized in this manner) do not spend more than your allowable time for any section finished or not, move on to the next section.

If you feel stress or tension, take a break...right now. Do not postpone a break when you feel too tense to work. A brief break for a minute or two...a trip to the water fountain or men's room...may mean the difference in passing. If the rules prohibit your leaving your seat, try closing your eyes and taking some long stretches in varying positions, pumping your muscles, rotating your head, etc.

TACTICS USED BY TEST WRITERS AND THE STRATEGY REQUIRED TO BEAT THEM

The scope, contents and style of an examination depends, of course, upon the writer or writers. All examinations are of limited scope i.e., an examination to text your knowledge or refrigeration is limited to refrigeration questions. There are only so many questions that can be composed on the subject without duplication. Variations on a theme may be diverse but the theme remains basic. Where problem solving is required as part of the exam, such as, solving the refrigeration load for a given set of conditions; the given conditions may vary greatly, but the basic steps for finding the solution remain unchanged.

Your examination may be written by a professional examination group, a college professor, a government official, a group of contractors acting as a "Board", or a registered engineer. Who writes the examination will determine the contents and style of questions; it is often possible to tell who wrote the exam by the way it reads. Government officials are inclined to load the examination with questions related to code, a contractor board with "pratical" type questions, and a college professor with "theoretical" type questions.

To avoid the possibility of fraud, or defraud, identical examinations are seldom, if ever, given twice. In those cases where an examination is given more than once, it alternates with other exams so that the same examination is never given in succession and will only reappear after 3 or 4 years. Usually the examiners will maintain a bank of questions for each category; the bank may contain several hundred or even thousand questions. When the examination is being prepared (say 50 questions are required for a particular exam) the questions are pulled from the bank at random. The chances are that some of the questions from an earlier examination will always reappear.

It is not unusual for different municipalities to obtain questions from a similar body, and in some cases, municipalities exchange examinations and exam information. Regardless of source or intent, all exams of a particular category will have some similar, if not identical questions. This is true because of the universality and--as mentioned above--the limited scope of any subject. You might be asked:

One ton of refrigeration is equal to 12,000 Btu/hr. True, or false: or;

$I = \frac{W}{E}$ True, or false?

These are universal truths, and both of these questions will appear, as shown, or in some variation, in any examination in the country. The chances are that some of the questions--or variations of them--that appear in the SAMPLE EXAM, will appear on your actual examination.

WHEN YOU GET THE SIGNAL TO BEGIN

RELAX: Read the directions carefully and be certain that you know what to do...and what not to do. You may be instructed not to write on your test booklet or question paper, but only on your answer sheet. You may be instructed not to write your name on any sheet, but only the number given you with your test. Follow instructions carefully; any violation could disqualify you.

Having read the directions and identified yourself in the proper spaces, put down your pencil and scan the entire exam. This overview of the whole examination is an important part of your strategy; it gives you perspective. The perspective overview introduces you to the style of the exam, it familiarizes you with the general tone and cadence of the exam. You may detect several styles in different sections, indicating different writers. You will have a "feel" that it's an easy or tough exam, that you are well prepared through study, or that many questions on control wiring (your weakness) appear. Further, a good overview often affords a clue to early questions in the exam through some revelations which appear further along. And, knowing how many questions there are will give you a fair indication of how much time you can spend per question..now you can pace yourself!

THE ESSAY QUESTION

An essay question requires you to compose an answer and write it out. Most modern examinations do not use this type of question because it can not be answered for machine-scoring; it must be read by the examiner and graded accordingly.

An essay type question requires more than a knowledge of your subject--you need to be able to express yourself, or explain that knowledge to others. Before answering, make certain that you are expressing your thoughts correctly. Examine the phrase in your mind before you set it down on paper; there may be a better way to express your thought, or perhaps the word position should be changed. For example, you write:

> "The man ran wildly down the street after the dog in red pajamas".

you meant:

> "The man in red pajamas ran wildly down the street after the dog".

or:

> "He is the man in the blue car's brother".

you meant:

> "He is the brother of the man in the blue car".

Watch out for the "multiple answer trap". An essay question may have more than one answer. Write all the answers which may apply; the examiner could be looking for the one you omitted. For example:

Q. One or more units of a system are not heating. What is the probable cause?

Think of *all* the likely causes and include them in your answer:

A. The air eliminators are not opening...steam trap is stuck...dirt in the lines...inoperative valve...failure in the control system...pocket in the steam line.

Another example:

Q. Why is the suction riser in a refrigeration system usually of smaller diameter than the horizontal lines?

Try to picture the circuit in your mind and answer the question fully: half an answer will be marked wrong.

A. To increase the velocity of the gas and insure the return of oil to the compressor.

Either half of the answer is correct, but the examiner would probably consider it incomplete...don't chance it.

THE COMPLETION, or FILL-IN QUESTION

This is a suggestive type question; some persons are more responsive to this type of question than others. The answer usually comes to mind quickly. If it doesn't, don't fret long; move on to the next questions and come back to that one later. A fresh look at it for the second time may suggest the answer. Example:

Q. The secondary high voltage windings of a small power transformer are color coded ___________

You would be expected to write in the word "red" into the blank space. If you do not know color coding, do not waste any time with that question; your experience will tell you that it is either black, green, white or red...take your guess and move along.

The fill-in question is not frequently used in modern exams, but you may encounter it.

THE MULTIPLE CHOICE QUESTION

This is the most popular type question, the kind you are most likely to encounter on your next examination. Multiple choice questions are written for most civil service exams, college entrance exams, municipal, county and state board exams, vocational training courses, and regular college examinations.

Although "trick" or "catch" questions will rarely appear on your exam, the questions may be framed in such a way as to throw you off guard. You are offered a cluster of either 4 or 5 answers to each question and you must select only one answer. Very often all the answers seem possible; you must select the most correct answer. When a skilled examiner has written the examination, the multiple choice question can be a real brain wringer. Do not hurry...examine each question carefully, regardless of how simple the answer looks at first glance...the obvious looking answer is often placed there just to force you to think through all the possible answers.

Attacking the Multiple Choice Question:

1. Look for the qualifying word in the question e.g.
   always...never least...most often...rarely
   most likely...least likely sometimes...always
   smallest...largest probably...possibly
   maximum...minimum best, most advisable,
   usually, greatest, chiefly, etc.

Any of these qualifiers offer a clue to the correct answer.

2. Don't assume the meaning of a word. The use of the abbreviation, e.g., above, was deliberately used to check your comprehension. If you do not know what e.g. means look it up in your dictionary. You should find it in the back section, "Abbreviations". If your dictionary does not carry such a list of abbreviations, it is unsuitable...get a good one. The meaning of words is critical to any exam; if you are unfamiliar with any words used in this study course or Sample Test, use your dictionary. As part of your study curriculum you should have a good *Glossary of Trade Terms.*

3. If you are unsure of the answer, apply the principle of elimination. That is, first eliminate the choices that you know are wrong; this narrows your choice down to 3 or perhaps 2 possible answers. Sometimes, the process of elimination will lead you directly into the correct choice.

4. Never allow yourself to be influenced by an answer "pattern". There is no pattern of answers. When answers appear to be following a certain pattern, it is usually coincidental. If the questions are deliberately placed in pattern...watch out! You may get a series of

questions like:

Q. Bare wire ends are spliced by the

a. Western Union method
b. rat-tail joint method
c. fixture joint method
d. all of the above

... in 3 or 4 successive questions the answer will be "d", as in the above. Then you will get:

Q. How are solderless connectors installed on conductors?

a. crimped on
b. lugged on
c. bolted on
d. all of the above

... the "d" pattern is broken, when you least suspect it!

In the following example you see an actual question which has appeared on many exams and is usually answered wrong. Most candidates would know the correct answer if they would think about it for a moment, but the question is phrased in such a way as to appear simple:

Q. A rotating vane anemometer is an indicating instrument that registers an air current in

a. mph
b. cfm
c. fpm
d. feet

Over ninety percent of examinees will mark "c"; the correct answer is "d". Do not rush into an answer, think for a moment before you make your selection.

THE MACHINE SCORED ANSWER SHEET

Figure 1.2 is a reproduction of an actual Test Answer Sheet frequently used for contractor examinations. Where this system is used, you will be given a Test Answer Sheet along with your examination papers. Each question will be numbered and you will be required to mark the corresponding number on the Answer Sheet.

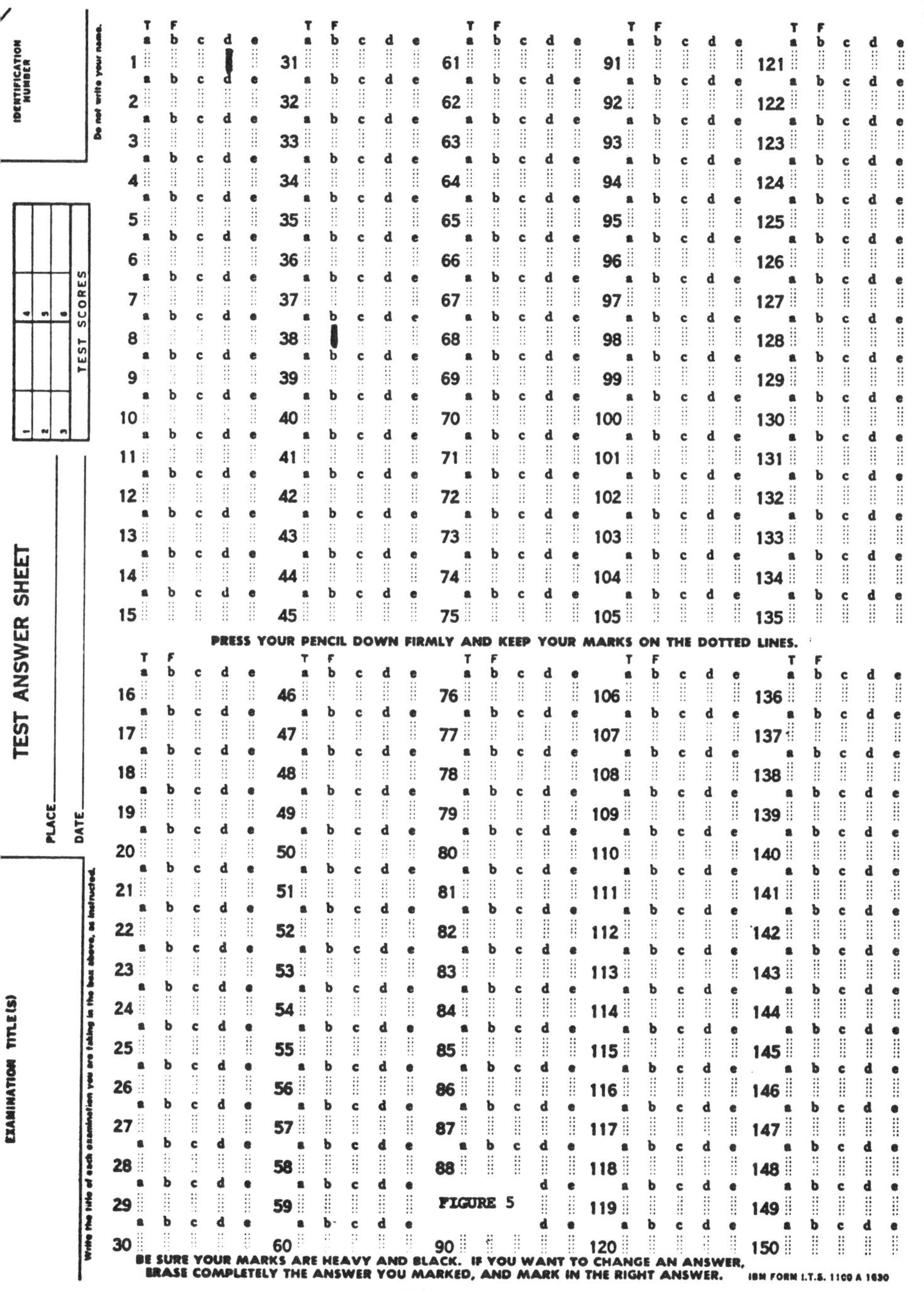
TEST ANSWER SHEET

IDENTIFICATION NUMBER

Do not write your name.

TEST SCORES

PLACE

DATE

EXAMINATION TITLE(S)

Write the title of each examination you are taking in the box above, as instructed.

PRESS YOUR PENCIL DOWN FIRMLY AND KEEP YOUR MARKS ON THE DOTTED LINES.

FIGURE 5

BE SURE YOUR MARKS ARE HEAVY AND BLACK. IF YOU WANT TO CHANGE AN ANSWER, ERASE COMPLETELY THE ANSWER YOU MARKED, AND MARK IN THE RIGHT ANSWER.

IBM FORM I.T.S. 1100 A 1630

Figure 1.2

Assume the first question on the examination is a multiple-choice:

1. Dade is a county in the state of:

    a. California
    b. Virginia
    c. New York
    d. Florida

The answer would be marked alongside 1.d., as shown in Figure 1.5

Assume question number 38 on the examination is a true-false:

38. It is not required to have a certificate of competency to perform building contracting work in Dade County.

The answer would be marked alongside 38.b., as shown in Figure 1.2.

If you examine the sample Test Answer Sheet, you will note that the spaces for answers are marked, a, b, c, d, e, or T F. Test Answer Sheets may vary slightly; before you commence your examination take a few moments to study the Sheet and make certain that you follow the sequence of numbering as well as any other instructions that may appear.

SUMMARY

1. Mark your STUDY SCHEDULE up to the maximum time your can afford, and stick to it. Discipline... discipline...discipline...

2. Memorize the 10 RULES FOR EFFICIENT STUDY and make them your way of life until you have secured your certificate.

3. Do not attempt to memorize answers to specific questions. It is not realistic to expect to know everything there is to know about your trade, craft, or profession, and it is no loss of pride to miss some answers. The main objective is to get a passing grade. Part of the strategy of getting a passing grade and receiving your certificate of competency, is understanding this objective and knowing your limitations. Remember, there are many licensed persons who know less than you do.

4. Concentrate your study on those areas in which you know you are weak, and strengthen your "I.Q." around those areas.

5. Knowing where to find the information is half the battle won. Familiarize yourself thoroughly with the table of contents and index of the recommended study material; this is essential for the "open book" section of your examination.

6. Do not study irrelevant material. Your recommended reference list may include some books from which only one chapter--or a short passage--is applicable. For example, if you are taking the examination for well drilling, you will find the book, *Plumbing* by Babbitt, listed. This does not mean that you are responsible for the entire text; confine your studying to those passages that apply to well drilling and piping only. Where volumes of the NFPA are listed, mark out those pamphlets cited, and the important sections referring to your craft, and ignore the remainder.

300 QUESTIONS

All of the following questions were taken from actual examinations for *Journeyman General Mechanical*. They have been selected on the basis of frequency of appearance in various examinations around the country; New York, Miami, Ft. Lauderdale, West Palm Beach, Los Angeles, Chicago, Detroit, are all represented. In some cases the questions were taken from Civil Service Refrigeration License examinations.

Although some effort has been made to keep the questions grouped by category, there is, of course, a percentage of crossover. Questions from refrigeration exams often appear on pressure piping exams and vice versa.

The different styles in writing and organization of material indicates that various persons worked in preparing the questions. For the most part the questions appear exactly as they were on the examinations. In a few cases the word order was changed to make them more understandable, and occassionally, spelling had to be corrected and symbols changed.

The candidate is expected to select the answer closest to the truth although that may not be the *exact* answer. Sometimes questions are poorly prepared or unfairly presented, and *sometimes questions are incorrectly graded.* If you fail an examination by only a few points and the law entitles you to a review - ask for it. *Know your rights!*

Answers to these questions will be found on page 241

1. The capacity of an evaporative condenser depends on what entering air temperature?

   (a) Dry bulb (c) Warm

   (b) Wet bulb (d) Differential

2. An expansion valve set for high superheat will cause:

   (a) Flooded evaporator (c) High back pressure

   (b) Starved evaporator (d) High head pressure

3. Open motors are generally designed to run a temperature rise 40C (72°F). What does this mean?

   (a) Indicates the temperature rise over the surrounding air of a motor when running at full name-plate conditions

   (b) Indicates the temperature rise of the motor

   (c) Indicates the temperature drop of the surrounding air

   (d) Indicates the temperature rise over the surrounding air or a motor when not running

4. What factors determine the size of an expansion tank?

   (a) The amount of space the water in the system requires in its expanded state

   (b) The amount of space the air in the system requires in its expanded state

   (c) The amount of air in system

   (d) The operating pressure in system

5. To check speeds or rotary equipment, which of the following instruments are used?

   (a) Tachometer (c) Psychrometer

(b) Manometer

(d) Velometer

6. The correct chemical symbol for Refrigerant-12 is:

(a) CCL_2F

(c) C_2CL_2F

(b) CCL_2F_2

(d) $CHCLF_2$

7. What is the gross weight of refrigerant cylinder?

(a) The weight of the refrigerant plus the weight of the cylinder

(b) The weight of the cylinder

(c) The weight of the contents only

(d) The weight of cylinder and line

8. If a thermostatic expansion valve is only partially feeding a coil, it is said to be:

(a) Operating at a low superheat

(b) Operating with no superheat

(c) Superheated

(d) Operating at a high superheat

9. On an A/C System, low power factor to a motor will cause it to:

(a) Not run at all

(c) Heat up

(b) Run slow

(d) Hum

10. A modulation thermostat employs what to achieve its purpose?

(a) Pressure

(c) Switch contacts

(b) A transformer

(d) A potentiometer winding

11. In a direct expansion chiller the refrigerant is on what side of the tubes?

(a) On the shell

(c) Outside

(b) Inside (d) Both sides

12. In reference to a hydronic system there are three rules needed to divert the air in the system to an expansion tank. Which one of these rules are not true?

    (a) We should be at the point of lowest pressure

    (b) We should be at the point of highest pressure

    (c) We should be at the point of highest temperature

    (d) We should be at the point of lowest velocity

13. In brine systems on measuring the pH reading of the brine, what pH would represent neutrality?

    (a) 7.0 (c) 5.0

    (b) 10.0 (d) 0

14. Too much refrigerant, that is, an overcharge, reacts differently with different refrigerant controls. Therefore, when the indication is overflowing and flood back, what type of refrigerant control is being used?

    (a) Thermostatic expansion valve

    (b) Automatic expansion valve

    (c) Low side float

    (d) High side float

15. Soft soldered joints on refrigeration piping which leak may be repaired by:

    (a) Cleaning the exposed area, refluxing and resoldering

    (b) Chipped and resoldered

    (c) Brazing

    (d) Being disassembled, cleaned, refluxed and resoldered

16. If you used Refrigerant-12 in a machine designed for Refrigerant-22, what will happen?

    (a) It will cause copper plating

    (b) Capacity of the machine will be greatly overloaded

    (c) Capacity of the machine will be greatly reduced

    (d) There will be no difference in capacity

17. A piping system in which the heating or cooling medium from several heat transfer units is returned along paths arranged so that all circuits composing the system or composing a major subdivision of it are practically equal length, is known as a:

    (a) Reversed return
    (c) Two pipe

    (b) Indirect
    (d) Down feed

18. High pressure drop in the liquid line will result in:

    (a) Increased efficiency

    (b) Increased capacity

    (c) Flash gas in the liquid line

    (d) Superheat in the liquid line

19. On a single temperature pneumatic control temperature system a pressure reducing valve reduces the tank pressure to maintain a main air pressure of how much?

    (a) 10 psi
    (c) 15 psi

    (b) 13 psi
    (d) 17 psi

20 When the thermo-controls are made of metal, the elements in the controls depend on the principle of:

    (a) Specific heat of the metal

    (b) Density of the metal

(c) Equal expansion of metals

(d) Unequal expansion of metals

21. It is easier to burst tubes in a flooded chiller than in a dry chiller in a freeze up.

(a) True (b) False

22. Freeze up in a flooded chiller is prevented by_________control.

(a) Low water (c) Suction pressure

(b) Low freon (d) Overload

23. If the float ball in a high side float becomes full of liquid, the float will:

(a) Open (c) Expand

(b) Chatter (d) Close

24. In a Dry Chiller, the freon is in the shell and the water is in the tubes.

(a) True (b) False

25. A cold equalizing line on an expansion valve indicates what fault with the valve?

(a) High superheat

(b) Leaking pin packing

(c) Low superheat

(d) Weak power assembly

26. An air conditioning unit discharges 360 fpm through a grill measuring 6 inches by 12 inches. What is the cfm?

(a) 180 cfm (c) 90 cfm

(b) 360 cfm (d) 720 cfm

27. A water cooled condensing unit is running high head pressure and has a water TD of 22 degrees. What is the probable cause of the high head?

(a) Dirty condenser

(b) Air in system

(c) Too much refrigerant

(d) Not enough water

28. How much condenser water is required for a five ton air conditioning system using tower water?

(a) 15 gpm
(c) 30 gpm

(b) 10 gpm
(d) 8 gpm

29. What should be the maximum time delay between the steps of an increment start motor?

(a) 15 seconds
(c) 30 seconds

(b) 5 seconds
(d) 1 minute

30. Electrical resistance is always measured in:

(a) Coulombs
(c) Ohms

(b) Henrys
(d) Watts

31. The oil safety swtch is operated by:

(a) Oil pressure

(b) The sum of oil pressure and crankcase pressure

(c) The difference between crankcase pressure and oil pressure

(d) The difference between head pressure and oil pressure

32. In the refrigeration system, what is the basic formula used in making a condenser analysis?

(a) Condenser temperature split = air inlet dry bulb temp - condenser temperature.

(b) Condenser temperature split = condensing temperature - air inlet dry bulb temperature

(c) Condenser temperature split = condenser pressure - air inlet dry bulb temperature

(d) Condenser temperature split = condenser temperature - air inlet wet bulb temperature.

33. What is the basic formula used in making an evaporator analysis?

(a) Evaporator temperature split = inlet wet bulb temp - evaporator temperature

(b) Evaporator temperature split = inlet air dry bulb temperature - evaporator temperature

(c) Evaporator temperature split = outlet air dry bulb temperature - evaporator temperature

(d) Evaporator temperature split = outlet air wet bulb temperature - evaporator temperature

34. What is the formula for analyzing the performance of the expansion valve?

(a) Valve superheat = temperature at TXV bulb - evaporator pressure

(b) Valve superheat = pressure at TXV bulb - evaporator pressure

(c) Valve superheat = temperature at TXV bulb - head pressure

(d) Valve superheat = temperature at TXV bulb - evaporator temperature

35. Refrigerant = Refrigerant-22
Inlet air dry bulb temperature = 65°F
Compressor outlet pressure = 165 psig
Condenser temperature split = ________?

(a) 53° F

(b) 49° F

(c) 23° F

(d) 16° F

36. "Condensing temperature *minus* leaving refrigerant temperature". The above temperature difference is used to check for what abnormal condition in an air conditioning or refrigeration system?

    (a) Refrigerant overcharge

    (b) Refrigerant undercharge

    (c) Dirty condenser

    (d) Any of the above

37. On an isolated condenser if there is no air in the system, the temperature corresponding to the pressure difference between the refrigerant and the air will finally be__________.

    (a) Over 10° F   (c) 8° - 9° F

    (b) 2° - 3° F or less   (d) The same or less

38. Stabilizing temperature conditions in an "air in system check" requires the condenser fan to be___________.

    (a) Induced draft   (c) Switched off

    (b) Either on or off   (d) Left on

39. In a refrigeration system, suction line pressure drop = refrigerant pressure in the evaporator minus____________?

    (a) Suction pressure at the compressor

    (b) 2 psi

    (c) Suction pressure at the evaporator

    (d) Superheat

40. Type of refrigerant - R-12
    Compressor suction pressure = 46.7 psig
    Inlet air dry bulb temperature = 75° F
    Evaporating temperature = _____° F

    (a) 25° F   (c) 10° F

    (b) 40° F   (d) 50° F

41. In the case of the previous question, evaporator temperature split = _______?

    (a) 50° F (c) 30° F

    (b) 20° F (d) 25° F

42. A refrigeration air conditioning service man finds that the suction line pressure drop on a system is 4 psig per 100 feet of line, which of the following would not be a cause?

    (a) Clogged suction line filter

    (b) Excessive oil in suction line

    (c) Dirty air filters

    (d) Pinched line

43. What other component analysis in an air conditioning system besides the evaporator and condenser would indicate low refrigerant charge?

    (a) Expansion valve (c) Discharge pressure

    (b) Suction pressure (d) All of the above

44. Complete the following readings: Refrigerant-12; Temperature at TXV bulb 55° F; suction gage pressure - 30 psig. Evaporation temperature_______?

    (a) 36° F (c) 32° F

    (b) 40° F (d) 15° F

45. Complete the following readings: Refrigerant-22
    Temperature at TXV bulb - 50° F.
    Suction gage pressure 69 psig
    Evaporation temperature 40° F
    Superheat_________?

    (a) 10° F (c) 12° F

    (b) 5° F (d) 15° F

46. Normal suction line pressure drop in a refrigeration system is estimated as:

    (a) 2 psig per 50 feet of line

(b) 4 psig per 100 feet of line

(c) 2 psig per 100 feet of line

(d) 4 psig per 50 feet of line

47. Complete the following reading from an air conditioning system. Temperature at TX valve bulb 50° F
Evaporating temperature 45° F
Valve superheat________?

(a) 10° F (c) 15° F

(b) 5° F (d) 41.7 psig

48. A refrigeration air conditioning service man in the course of a TX valve analysis feels the liquid line downstream and upstream of the solenoid valve or strainer, what measurement would lead him to do this?

(a) High superheat (c) High back pressure

(b) Low superheat (d) Low suction temperature

49. Liquid Refrigerant-22 enters an evaporator @ 50.2 psig 20° F and leaves @ 38° F. Its pressure as it leaves will be___________.

(a) 36.9 psig (c) 50.2 psig

(b) 43.3 psig (d) 66.1 psig

50. Flooded feeds are different from direct expansion feeds in that they_____________.

(a) Are not as efficent

(b) Use bigger evaporators

(c) Allow for regurgitation of liquid refrigerant

(d) Allow for recirculation of liquid refrigerant

51. The hi-side float is controlled by_____________.

(a) The rate of refrigerant evaporation

(b) The rate of refrigerant condensation

(c) The evaporator temperature

(d) The evaporator pressure

52. The lo-side float is controlled by__________.

(a) The rate of refrigerant condensation

(b) The rate of refrigerant evaporation

(c) The evaporator temperature

(d) The evaporator pressure

53. What are the suggested possibilities of trouble indicated by the gauge and superheat readings, low suction pressure - low superheat?

(a) Overcharge of refrigerant

(b) Improper superheat adjustment

(c) Compressor undersized

(d) Evaporator oil-logged

54. A reverse acting pneumatic humidistat increases the air pressure to a controlled device when the humidity____________.

(a) Increases

(b) Decreases

(c) Remains static

(d) Mixes

55. If a drum of dichlorodifluoromethane; partially filled with liquid refrigerant, is stored in a 32° room the pressure in the drum would be:

(a) 10 lb

(b) 124 lb

(c) 16 lb

(d) 30 lb

56. When the pressure difference across a capillary tube is increased ____________.

(a) The flow of refrigerant increases

(b) The flow of refrigerant decreases

(c) It gets noisy

(d) There is no change in flow rate

57. Another name for a normally open valve that is pneumatically controlled:

(a) Indirect-acting valve

(b) Diverting valve

(c) Reverse acting valve

(d) Direct-acting valve

58. By using nitrogen within the pipe during brazing operation:

(a) The explosion hazard is confined

(b) Discoloration is prevented

(c) Copper oxide formation is prevented

(d) Exterior scale formation is minimized

59. When using nitrogen within the pipe during the brazing operation, the nitrogen pressure should be:

(a) Very low

(b) 50-100 psig

(c) 250 to 300 psig

(d) Slightly below atmospheric

60. "Bullheaded Tees" are not acceptable in discharge lines.

(a) True

(b) False

61. When individual suction lines are taken from a common suction header, the lines should not be taken from the ____________ of the header.

(a) Top

(b) Side

(c) Bottom

(d) End

62. The latent head used in changing a liquid to a gas or vapor is called the latent heat of:

(a) Fusion
(b) Pressurization
(c) Condensation
(d) Vaporization

63. Speed of a belt driven compressor may be increased by:

(a) Increasing pulley size on motor
(b) Increasing pulley size on compressor
(c) Using belt dressing on pulleys
(d) Decreasing pulley size on motor

64. A Journeyman uses 53 pieces of pipe each 6 feet long and 8 pieces each 12 feet long. How many feet of pipe did he use?

(a) 515 feet
(b) 414 feet
(c) 314 feet
(d) 400 feet

65. On a horizontal shell and tube condenser using a water regulating valve, where should the valve be placed?

(a) On the inlet side of the condenser
(b) Either side of the condenser
(c) On both sides of the condenser
(d) On the oulet side of the condenser

66. What is the maximum percent of the volume of a refrigerant receiver that can be used for storage capacity?

(a) 20%
(b) 50%
(c) 80%
(d) 100%

67. What happens when a capillary tube system is overcharged?

(a) It has a high head pressure

(b) It will partially defrost

(c) It will not sweat and frost back

(d) It will have below normal low side pressure

68. An air compartment or chamber to which one or more ducts are connected and which forms part of either the supply of return system is called:

(a) A proscenium
(b) A vortex
(c) A plenum
(d) A plectrum

69. What does wet bulb depression mean?

(a) A cavity in the bulb

(b) The wet bulb is located below the dry bulb

(c) The difference in reading between the wet bult thermometer and the dry bulb thermometer

(d) The inaccuracy of the thermometer

70. What does the inner tube opening register on a pitot tube?

(a) Total pressure
(b) Velocity pressure
(c) Static pressure
(d) Total pressure minus velocity pressure

71. With what basic function of air conditioning are duct velocities most concerned?

(a) Air heating
(b) Air cooling
(c) Air humidifying
(d) Air distribution

72. Do all filters have a pressure drop?

(a) Yes
(b) No
(c) Only if they are dirty
(d) Only if high air speeds are used

73. The temperature of the refrigerant in the generator of absorption system is:

(a) Greater than that of the absorbent

(b) Less than that of the absorbent

(c) The same as that of the absorbent

(d) Greater and less than that of the absorbent

74. An instrument for measuring pressure, essentially a U-Tube partially filled with a liquid, usually water mercury, or a light oil, so constructed that the amount of displacement of the liquid indicates the pressure being exerted on the instrument:

(a) Potentiometer (c) Manometer

(b) Volometer (d) Anemometer

75. In practically all coils used for cooling air, the flow of air and water through the coil are in the____________direction to each other.

(a) Water flow (c) Same

(b) Parallel (d) Opposite

76. The take-off (end to center measurement) of a standard long radius 90° weld ell is equal to:

(a) The diameter

(b) Twice the diameter

(c) 1-1/2 times the diameter

(d) 1/2 the diameter

77. An important factor in the piping of supply and return fuel oil lines to an underground oil storage tank is that______ ________shall always be used.

(a) Swing joints (c) Two pipes

(b) Clean pipe (d) One elbow

78. A standard 6" thread that has been properly cut will allow the fitting to hand tighten________

times.

(a) 3 (c) 8

(b) 6 (d) 1-1/2

79. All branch (take-off) lines from a steam main should be taken from the__________.

(a) Side (c) Bottom

(b) Top (d) Top or bottom

80. Where oil fuel tanks are lower than the burner, on fuel burning equipment, it is recommended that they have what kind of piping system?

(a) Oversize (c) 2 pipe

(b) 3 pipe (d) 1 pipe

81. An air control system consists of a boiler fitting and tank fitting. The combined function of the two fittings is:

(a) To relieve air to the atmosphere

(b) To control the high pressure air cut-out

(c) To remove air from the heating system at a point of low velocity and put that air in the compression tank

(d) To remove the air from the compression tank and add it to the heating system

82. The gas conversion burner which uses a small blower and discharges the flame horizontally into the fire box is an____________.

(a) Upshot atmospheric burner

(b) Inshot atmospheric burner

(c) Upshot power burner

(d) Inshot power burner

83. Pressure tanks in forced flow hot water heating systems should contain_____________.

(a) Water only (c) Steam only

(b) Air only (d) Water and air

84. Two monoflow fittings for radiation below the main on a one-pipe forced flow hot water heating system are necessary in order to overcome the tendency for____________.

(a) Water to seek the lowest level

(b) Hot water to rise

(c) Warm air to rise

(d) None of these

85. Steam fittings shall *not* be of what material?

(a) Cast iron (c) Malleable iron

(b) Galvanized iron (d) Brass

86. The purpose of a steam trap is to:

(a) Trap steam and discharge it to atmosphere

(b) Prevent discharge of steam into the return piping and to eliminate air and condensate from the steam mains

(c) Prevent discharge of air into the return piping and to eliminate steam and condensate from the steam mains

(d) To collect air

87. What type boiler has water around the outside of the tubes being heated by the hot gases within the tubes?

(a) Water tube boiler (c) Radiant boiler

(b) Fire tube boiler (d) Atmospheric boiler

88. The science of heating and cooling with liquids is called:

(a) Hydrostatics (c) Hydronics

(b) Saturation (d) Solar

89. The purpose of a Hartford Loop is to:

(a) Prevent backing water out of a boiler

(b) Remove air from the return line

(c) Allow for pipe expansion

(d) Provide a balance for the steam header

90. On a steam system the operation of an inverted bucket trap is based on the:

(a) Rise and fall of a float

(b) Combination of steam pressure and weight of the condensate

(c) Steam pressure drop

(d) Weight of the water in the trap

91. In a forced flow hot water heating system, when piping must be run around an obstacle, such as a beam, what is advisable to do?

(a) Drop the pipe below the beam

(b) Loop over the beam only

(c) Install steam trap at low point

(d) Reroute pipe away from beam

92. The low voltage controls in a gas fired furnace are wired up in:

(a) Series connection (c) #8 wire

(b) Parallel connection (d) #10 wire

93. On water pipes an eccentric reducer should be used on which of the following:

(a) Horizontal pipes (c) Vertical pipes

(b) Brass pipes (d) Wrought iron pipes

94. The three most common steam traps in use are the:

(a) Float thermostatic impulse type

(b) Float pea impulse

(c) Float thermostatic vacuum

(d) High side thermostatic impulse

95. All piping, valves, and fittings used to connect instruments to main piping, to other instruments or apparatus, or to measuring equipment is known as:

(a) Control Piping

(b) Instrument Piping

(c) Sampling piping

(d) Branch Piping

96. Air in oil line to burner may cause which one of the following:

(a) Pulsation

(b) Excessive fuel consumption

(c) Blue flame

(d) Dirty nozzle

97. Condensate may be formed on fire side of a gas boiler if what condition exists?

(a) Not enough insulation on boiler

(b) Draft control not properly set

(c) Fuel-air ration not correct

(d) Water temperature maintained too low in the boiler

98. An ASME pressure relief valve on a boiler is usually provided with a hand lever. Its primary purpose is to:

(a) Close valve once it has opened

(b) Change operating pressure

(c) Test valve and relief of air on initial fill

(d) Provide manual safety

99. Open expansion tanks are open to atmosphere and are located:

    (a) Anywhere in piping system

    (b) Discharge side of pump

    (c) Suction side of pump obove highest unit in system

    (d) At heating-cooling change over valve

100. In the discussion of oil heating using a fuel pump, the number of pounds per square inch the pump pump pressure must drop to close nozel valve is known as:

    (a) Delivery

    (b) Valve differential

    (c) Head of oil

    (d) Lift

101. A two-stage refrigeration system should be used when:

    (a) Ammonia is used as the refrigerant

    (b) Compression ratio would be very high

    (c) Compression ratio would be very low

    (d) Suction temperature would be very high

102. On refrigeration, what type pipe or tubing is most commonly used with ammonia?

    (a) Iron or steel

    (b) Iron or brass

    (c) Copper or brass

    (d) Any of the above

103. What control would you use on a capillary tube or automatic expansion valve system to maintain desired temperature?

    (a) Dual pressure control

    (b) High pressure control

    (c) Temperature control

    (d) Low pressure control

104. The two basic types of evaporators are:

(a) Finned and plate

(b) Prime surface and finned

(c) Direct expansion feed and plate

(d) Flooded and direct expansion feed

105. R-22 will support combustion under certain conditions.

(a) True (b) False

106. A refrigerant oil should maintain sufficient body to lubricate at high temperatures and yet be fluid enough to flow at low temperatures.

(a) True (b) False

107. Most refrigerant oils are paraffin based.

(a) True (b) False

108. Horizontal piping runs should:

(a) Pitch toward the compressor

(b) Pitch away from the compressor

(c) Pitch in direction of flow

(d) Be level

109. If the driver pully is three inches and the motor rpm is 1700, what is the diameter of the driven pully if the desired rpm is 850?

(a) 8.3 (c) 6

(b) 7 (d) 5

110. What type of threads are used on flare fittings?

(a) National fine (c) U.S.S.

(b) National course (d) Standard pipe threads

111. In a beverage cooler installation the sweet water bath is made of____________.

(a) Alcohol
(b) Alcohol and brine
(c) Tap water
(d) Sugar solution

112. A refrigeration system is operating with refrigerant-12 at 37 psig suction pressure and the temprature of the R-12 at the end of the evaporator at the location of the TX valve-feeler bulb is 50°F. What is the superheat reading?

(a) 10° F
(b) 5° F
(c) 15° F
(d) 0° F

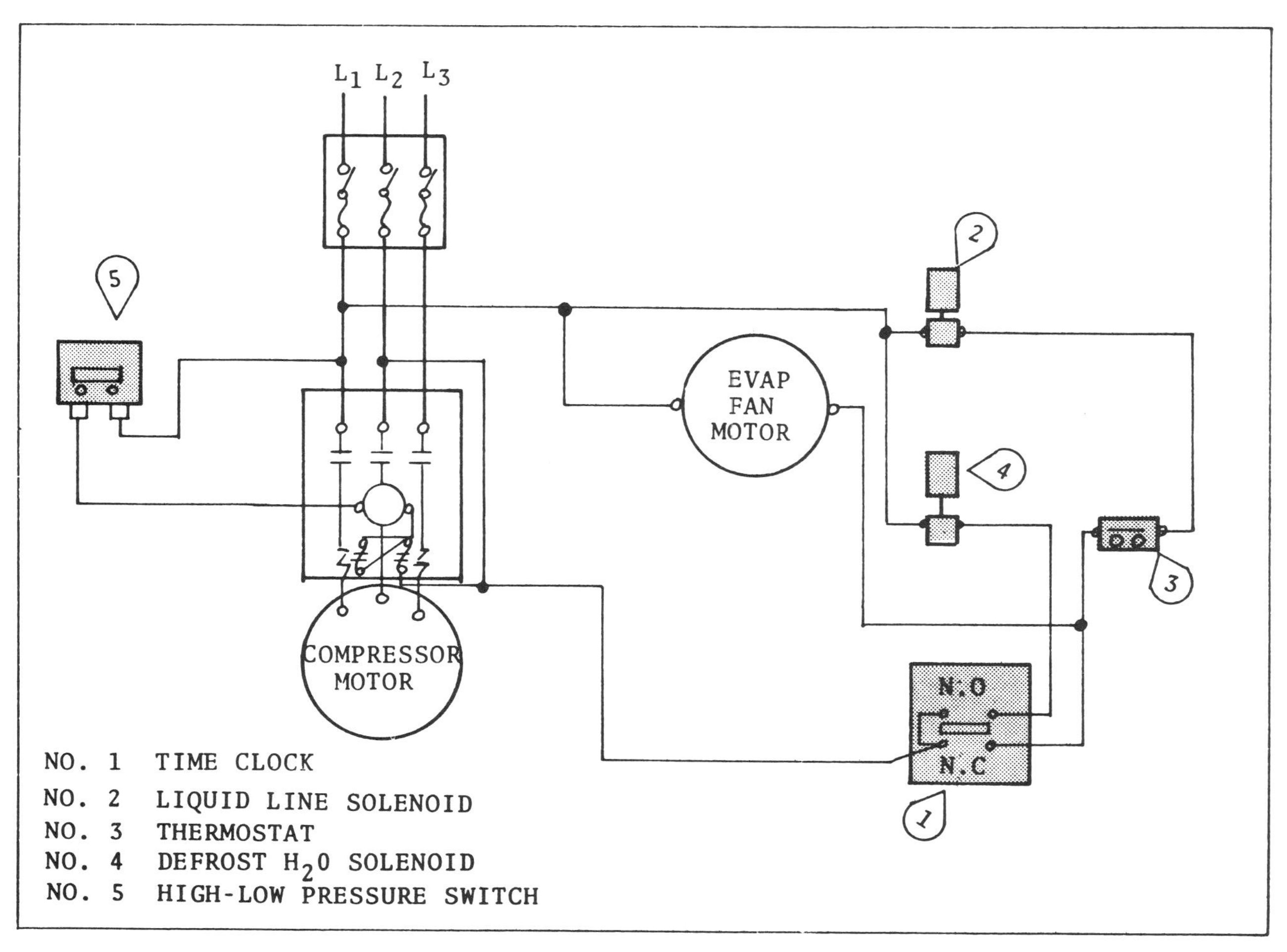
L1 L2 L3
EVAP
FAN
MOTOR
COMPRESSOR
MOTOR
N.O
N.C
1
2
3
4
5
NO. 1 TIME CLOCK
NO. 2 LIQUID LINE SOLENOID
NO. 3 THERMOSTAT
NO. 4 DEFROST H2O SOLENOID
NO. 5 HIGH-LOW PRESSURE SWITCH

FIGURE 1.3

IN RELATION TO THE REFRIGERATION CONTROL CIRCUIT SHOWN IN THE WIRING DIAGRAM IN FIGURE 1.3, ANSWER QUESTIONS 113 THROUGH 122, TRUE OR FALSE. IF TRUE, MARK A, IF FALSE, MARK B.

113. The entire control circuit is line voltage.

(a) True (b) False

114. When the time clock switch closes on the normally open position it will stop the fan motor while the compressor will continue to run.

(a) True (b) False

115. When the thermostat is satisfied it will immediately shut down the compressor.

(a) True (b) False

116. The sole function of the time clock is to allow defrosting of the coil.

(a) True (b) False

117. When the defrost cycle has been completed by predetermined time setting, the following takes place: The liquid line solenoid opens, the H_20 defrost solenoid closes, the evaporator fan motor starts up, and the time clock goes to its normally open position.

(a) True (b) False

118. The compressor motor safety is the hi-lo pressure control.

(a) True (b) False

119. If conditions make it necessary the compressor will recycle to pump down.

(a) True (b) False

120. If the thermostat were removed from the circuit, the evaporator fan motor would run continuously.

(a) True (b) False

121. Whenever the evaporator fan motor is not running the H_2O solenoid valve is wide open thus allowing the water to circulate across the evaporator coils.

(a) True (b) False

122. If the system pressure reaches a point below the setting of the hi-low pressure switch, the entire system becomes de-energized.

(a) True (b) False

123. What is the standard pressure control setting for a 36 degree walk-in cooler with F-12 on a defrost cycle?

(a) 20-40 (c) 15-35

(b) 10-30 (d) 0-32

124. A check valve on a multi-temperature system is located where and in what line?

(a) Coldest unit section

(b) Warmest unit section

(c) Either A or B

(d) Liquid line

125. What is the reason for the inlet water entering the bottom of a water cooled condensor?

(a) No reason (c) Counterflow to air

(b) Liquid sub-cooling (d) For air purging

126. In a Dry or Flooded chiller of the same capacity, which one has the largest refrigerant charge?

(a) Flooded (b) Dry

127. A suction pressure regulator cannot be used to regulate evaporator pressure.

(a) True (b) False

128. A refrigerant in mechanical refrigeration systems produces its greatest cooling effect __________.

(a) Through evaporation (c) By compression

(b) By condensing (d) By superheating

129. A valve in the liquid line usually located immediately at the condenser or receiver liquid outlet is a______________.

(a) Liquid shut-off valve

(b) Discharge shut-off valve

(c) Receiver shut-off valve

(d) Suction shut-off valve

130. The term "induced draft" could refer to:

(a) A type control

(b) A type of compressor

(c) A type of cooling tower

(d) A type of diagram

131. The temperature of the air surrounding the object under consideration is:

(a) Wet bulb temperature

(b) Condensing temperature

(c) Ambient temperature

(d) Dry bulb temperature

132. Where is the liquid receiver usually located in a capillary tube system?

(a) There is none

(b) In the cooling coil

(c) At the outlet of the condenser

(d) In the suction line

133. What may result if a too thick compressor cylinder head gasket is used?

(a) A knocking sound in the compressor

(b) A decrease in the volumetric efficiency on the compressor

(c) Oil pumping by the compressor

(d) No pumping

134. Where is a metering type two-temperature valve usually used?

(a) When a small temperature difference is desired

(b) When a large temperature difference is desired

(c) Any multiple system

(d) Only on display cases

135. With the compressor running, the pressure in a fully active evaporator using Refrigerant-12 is 21 lbs/sq. in gauge. The temperature of the main part of the evaporator is:

(a) 10° F (c) 0° F

(b) 20° F (d) 30° F

136. At what suction pressure is a condensing unit at its highest capacity in Btu's per hour, using Refrigerant-12?

(a) 5 psi gauge (c) 15 psi gauge

(b) 25 psi gauge (d) 40 psi gauge

137. On a pressure controller, which bellows controls the low pressure switch?

(a) Small bellows (c) Neither Bellows

(b) Large bellows (d) Both bellows

138. If three unmarked gas cylinders R-12, R-22 and R-500 respectively are all about one-half full and have been stored in the same room for several days at a temperature of 80° F, which cylinder contains the R-22?

(a) There is no way to tell if not marked

(b) The cylinder that gives a gage reading of 103#

(c) The cylinder that gives a gage reading of 84#

(d) The cylinder that gives a gage reading of 145#

139. The boiling point of ammonia refrigerant at atmospheric pressure is:

(a) 22° F below zero (c) 28° F below zero

(b) 75° F above zero (d) 41° F below zero

140. In a refrigeration system a solenoid valve is in the open position when the:

(a) Electricity is off

(b) Electricity is on

(c) Valve is under pressure

(d) Valve is not under pressure

141. One of the most useful tools for the serviceman is a gage manifold. It contains two shut-off valves and ____________ external connections.

(a) 3 (c) 2

(b) 5 (d) 4

142. For the ordinary installation what is considered good rule of thumb practice in spacing at intervals pipe hangers or supports for pipe:

   (a) 5 feet   (c) 10 feet

   (b) 20 feet   (d) 30 feet

143. On the installation of a pressure regulator for use in process work you should always install a relief valve and a pressure gage on the high pressure side of the regulator.

   (a) True   (b) False

144. Pop safety valve for steam air or gas should always be installed with the stem in a horizontal position.

   (a) True   (b) False

IN ANSWERING THE FOLLOWING TWO QUESTIONS IDENTIFY USE OF POSITIONS OF GAGE MANIFOLD FROM THIS DIAGRAM.

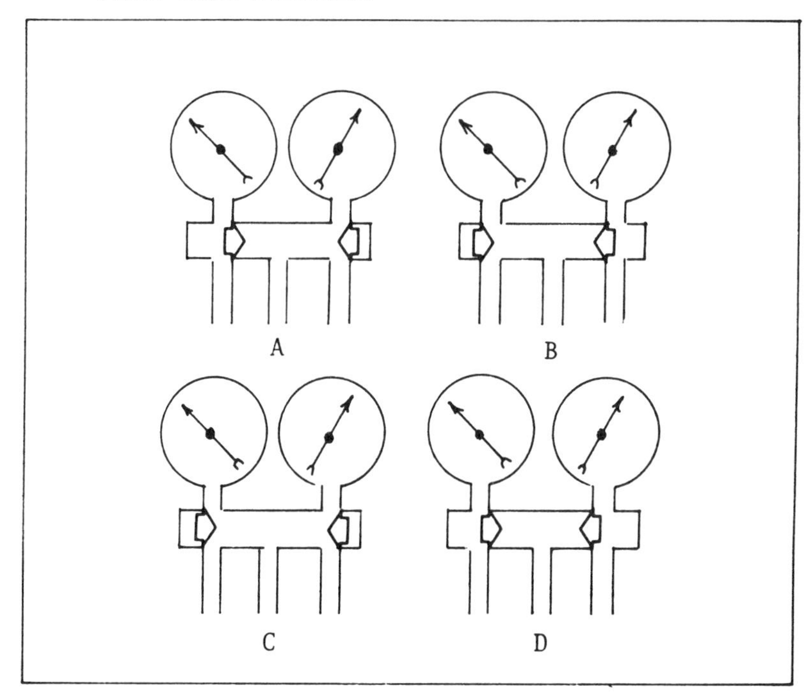

145. Gage manifold position for charging refrigerant or adding oil:

(a) A (c) C

(b) B (d) D

146. Gage manifold position for gauge readings:

(a) A (c) C

(b) B (d) D

147. Fuses used to prevent the use of a larger amperage rating fuse to correct an overload condition and are screwed into the plug fuse socket are known as:

(a) Fustats (c) Ferrule fuses

(b) Fustrons (d) Cartige fuses

148. A common single-phase motor overload protector, with a two-pin overload, sensitive to both temperature and current, usually located inside of motor terminal box and connected electrically in series with the motor's common terminal is:

(a) Overload relay (c) Fuse

(b) Circuit breaker (d) Klixon

149. On an electric motor insulation test you would perform a "polarity Index Test". This test requires what instrument?

(a) Ammeter (c) Megohmmeter

(b) Ohmmeter (d) Test Light

150. On motors less than 1 hp to determine if this motor has a ground; that is to see if any wire within a single-phase motor has broken or come in direct contact with the housing or shell causing a direct short to ground, and using an ohmmeter and taking a reading. This motor is probably grounded if the resistance is below ______________.

(a) 1,000 Ohms (c) 2,000,000 Ohms

(b) 230,000 Ohms (d) 1,000,000 Ohms

151. Which schedule pipe would have the greater wall thickness?

(a) Schedule 20 (c) Schedule 80

(b) Schedule 40 (d) Schedule 10

152. On piping systems, steel and wrought iron pipe joints may not be:

(a) Screwed (c) Flanged

(b) Welded (d) Soldered

153. On welded pipe fittings the center to face dimension of a standard 6" 45° weld ell is:

(a) 3-3/4" (c) 3"

(b) 5" (d) 6"

154. What instrument shall be installed on the low pressure side of a reducing valve?

(a) Ampmeter (c) Pressure gauge

(b) Flow meter (d) Pressure control

155. It is not essential that a check valve be installed on City Water Lines connected to heating equipment.

(a) True (b) False

156. A non-filler metal electrode used in are welding consisting of a tungsten wire is known a:

(a) Brazing rod (c) Welding rod

(b) Tungsten electrode (d) Carbon electrode

Give "A" measurements for the following pipe sizes. "B" remains 16' 4-3/8". Figure 1.4

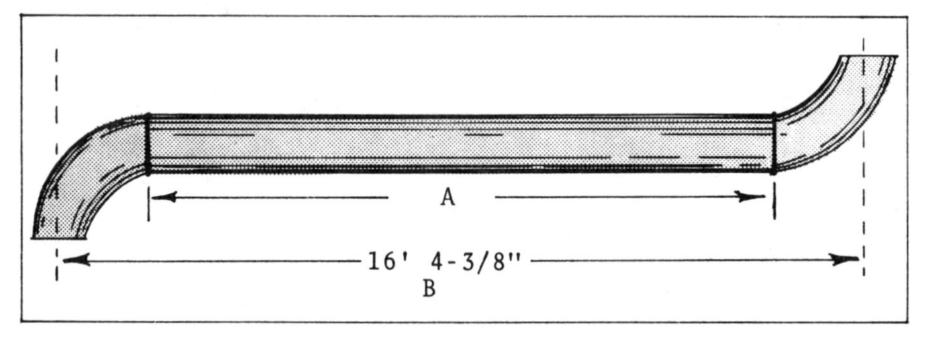

FIGURE 1.4

157. 2" Pipe - LR Butt Weld 90° Ells

(a) 16'0-3/8" (c) 15'10-3/8"

(b) 15'8-3/4"

158. 4" Pipe - LR Butt Weld 90° Ells

(a) 15'8-3/8" (c) 15'4-3/8"

(b) 16'6-3/8"

159. 8" Pipe - SR Butt Weld 90° Ells

(a) 15'0-3/8" (c) 14'8-3/8"

(b) 14'4-3/8"

160. 10" Pipe - SR Butt Weld 90° Ells

(a) 14'8-3/4" (c) 14'6-3/8"

(b) 13'10-3/8"

Note: To solve the problems above, see pages 160, 161, and 258

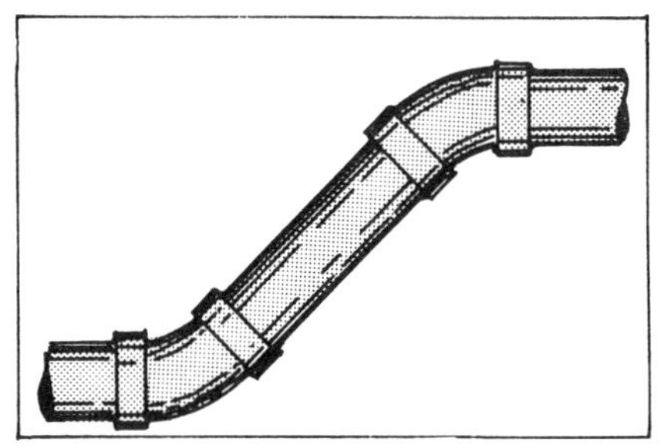

FIGURE 1.5

45° VERTICAL OFFSET

161. Set is 1'0" - What is the travel? (FIGURE 1.5)

    (a) 15'7-3/4" (c) 16-31/32"

    (b) 18-3/8"

162. Set is 1'6" - What is the run? (FIGURE 1.5)

    (a) 25-1/2" (c) 18"

    (b) 25-7/8"

163. Travel is 2'3" - What is the set? (FIGURE 1.5)

    (a) 19-1/16" (c) 3'2-1/2"

    (b) 27"

164. What is the OD of 4" Schedule 40 Pipe?

    (a) 4" (c) 4-7/8"

    (b) 4-1/2"

165. What is the OD of 4" Schedule 80 Pipe?

    (a) 4" (c) 4-7/8"

    (b) 4-1/2"

166. What is the OD of 16" standard weight pipe?

    (a) 16" (c) 16-3/4"

    (b) 16-5/8"

Note: To solve the problems above, *see pages 144, 145, 160, and 161*

167. OD of 2" pipe is 2.375 - What is the circumference?

(a) 7-7/16" (c) 7-11/16"

(b) 7-9/16"

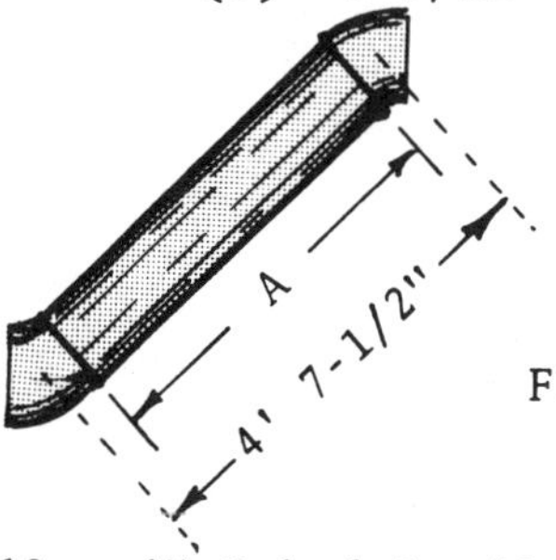

In this 45° offset problem, what is the A measurement for the pipe sizes listed in Questions 168, 169, 170?

FIGURE 1.6

168. 4" Schedule 40 pipe

(a) 4'2-1/2" (c) 3'7-1/2"

(b) 3'11-1/2"

169. 6" Schedule 80 pipe

(a) 4'0" (c) 3'1-1/2"

(b) 3'7-1/2"

170. 8" Schedule 160 pipe

(a) 3'9-1/2" (c) 2'7-1/2"

(b) 3'3-1/2"

171. In laying out a 3 piece 90° Ell, what would be the degree you would cut on the end of pipe?

(a) 15° (c) 45°

(b) 22-1/2°

172. What kind of thread does a street ell have?

(a) External threads only

(b) Internal threads only

(c) Both external and internal threads

(d) NF and NC threads

173. You should always use a full face gasket on a ________________.

(a) Ring type joint flange

(b) Raised face flange

(c) Flat face flange

174. What is considered to be the most important pass and the one that causes the most failures when testing on pipe?

(a) Weave (c) Root

(b) Lace (d) Cover

175. What tool is designed for use on pipe and screwed end fittings only:

(a) Monkey wrench (c) Open end wrench

(b) Pipe wrench (d) Chain pipe wrench

176. When the pitch of the thread on a pipe is 1/4" how many turns are required to thread 2-1/2" of the pipe?

(a) 8 (c) 12

(b) 10 (d) 13

177. Two pipes with an area of 3 square inches and 4 square inches respectively discharge into a single header. What is the diameter of the header if it has an area equal to the sum of the area of the two pipes?

(a) 4" (c) 3"

(b) 6" (d) 5"

178. A pipe run is 81.375 feet long. If divided into 21 equal parts, how long will each division be?

(a) 3.875 (c) 4.015

(b) 3.750 (d) 4.090

179. On flanged joints, bolts should be tightened in what manner?

(a) Cross over method (c) By hand only

(b) Rotation (d) Welded

180. All piping systems shall be capable of withstanding a hydrostatic test pressure of how many times the designed pressure?

(a) 3 (c) 2

(b) 1-1/2 (d) 2-1/2

181. How is stud constructed?

(a) With a hexagonal head

(b) With a round head

(c) With threads on both ends

(d) With a screwdriver head

182. The circumference of a circle 10" diameter is:

(a) 31.416" (c) 33"

(b) 32.614" (d) 33.416"

183. Resistance to flow in a piping system can be decreased most satisfactorily by:

(a) Increasing pipe size

(b) Increasing the pressure

(c) Increasing the velocity

(d) Decreasing the pipe size

184. What fitting should be used to connect 1/4" OD copper tubing flared to a 1/4" internal pipe thread opening in a water pump body?

(a) A union (c) A tee fitting

(b) A half union (d) A street ell

185. A pipe reducer where the center line of the larger pipe is out of line with the center of the smaller pipe is known as:

(a) Concentric reducer

(b) Out of line reducer

(c) Eccentric reducer

(d) Bell reducer

186. Copper tubing joints may be soldered, not brazed, when used in refrigerating systems containing Group ____ refrigerants.

(a) 3

(b) 2

(c) 2 and 3

(d) 1

187. Globe valves should always be installed with pressure on top of the disc with no exceptions.

(a) True

(b) False

188. Threaded and tapped fittings that are screwed into the end of other fittings or valves to reduce the size of the end openings are known as:

(a) Reducers

(b) Nipples

(c) Bushings

(d) Increasers

189. Pipe dope should be applied in what manner?

(a) Female threads only

(b) Male and female threads

(c) Male threads only

(d) Either male or female threads

190. On a district heating piping system, the after erection hydrostatic test shall be made with a test medium not in excess of ___ degrees, or pressure less than ____ psi.

(a) 212°F and 50 psi

(b) 200°F and 25 psi

(c) 50°F and 100 psi

(d) 100°F and 50 psi

191. All pipes should be ____________ after cutting to lengths.

(a) Cleaned (c) Checked

(b) Reamed (d) Threaded

192. What welding fumes are toxic and should be avoided?

(a) Galvanized (c) Stainless

(b) Cast Iron (d) Black steel

193. Which valve is not suitable for throttling flow?

(a) Globe (c) Check

(b) Gate (d) Swing

194. Flow through swing check valves are in a straight line and without restriction at the seat. This effect on flow is the reason for generally using swing checks in combination with __________ valves.

(a) Globe (c) All

(b) Lift check (d) Gate

195. If the gage pressure at the bottom of a 1/2" pipe filled one foot high with water is .434 psi the gage pressure at the bottom of a 2" pipe filled one foot high would be:

(a) .434 psi (c) .868 psi

(b) 1.302 psi (d) 1.736 psi

196. Figure 1.7 shows a 45° offset around a square obstruction; to find the starting point for the offset, find the distance of A if C = 15 in. and D = 8 in.

(a) 27-1/2" (c) 24-11/32"

(b) 28-9/16" (d) 26-10/32"

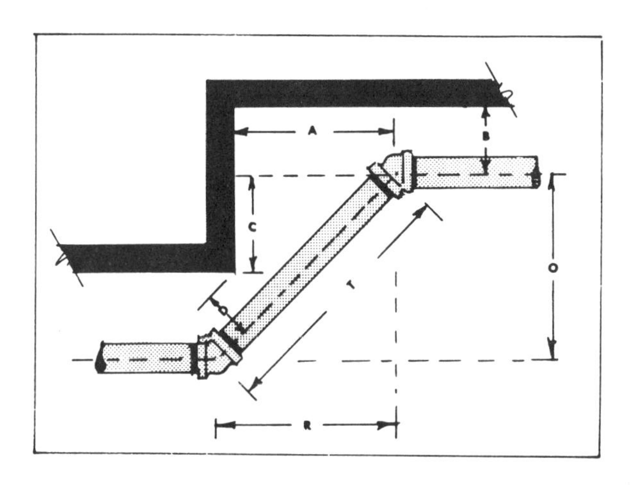

FIGURE 1.7

197. The allowable stress for seamless copper tubing is _______psi.

(a) 18,500 (c) 10,200

(b) 15,900 (d) 4,000

198. 124 pipe hangers weigh 100 lbs. How many hangers weigh 150 lbs?

(a) 166 (c) 186

(b) 206 (d) 176

199. The area of a square is found by:

(a) A = Width x height x length

(b) A = π x radius square

(c) A = the square of the hypotenuse

(d) A = Length x width

200. The Fahrenheit scale is based on boiling water having a sea level temperature of:

(a) 459 (c) 180

(b) 212 (d) 100

201. Hidden heat in refrigeration work is referred to as:

(a) Intensity of heat

(b) Latent heat

(c) Heat & thermometer can "sense"

(d) Cold

202. Absolute zero on the Fahrenheit scale equals:

(a) -459 (c) -100

(b) -273 (d) 0

203. The heat used to change a liquid to a gas or vapor is called latent heat of:

(a) Absorption (c) Fusion

(b) Vaporization (d) Liquid

204. A thermometer is said to "sense":

(a) Heat of fusion (c) Sensible heat

(b) Latent heat (d) Specific heat

205. Five pounds of water heated to raise the temperature (sea level) two degrees requires:

(a) 25 Btu's (c) 5 Btu's

(b) 10 Btu's (d) 15 Minutes

206. Superheat is heat added:

(a) In changing liquid to vapor

(b) In raising temperature of water

(c) After all liquid has been changed to vapor

(d) To increase pressure

207. The cooling component of a refrigeration system is called:

(a) A canteen
(b) An ice box
(c) An evaporator
(d) A conceiver

208. A refrigerant gives up heat when it:

(a) Condenses
(b) Evaporates
(c) Regurgitates
(d) Vaporizes

209. The thermostatic expansion valve is designed to maintain:

(a) A constant flow
(b) Constant temperature
(c) Constant superheat
(d) Constant pressure

210. A capillary is located:

(a) In the liquid line feeding the coil
(b) At the leaving side of the coil
(c) In the hot gas line
(d) At the condenser inlet

211. The automatic expansion valve maintains a constant pressure at:

(a) The condenser inlet
(b) The evaporator inlet
(c) The liquid line
(d) The condenser outlet

212. If a drum of dichlorodifluoromethane; partially filled with liquid refrigerant, is stored in a 32° room the pressure in the drum would be:

(a) 10 lbs. (c) 16 lbs.

(b) 124 lbs. (d) 30 lbs.

213. Condensing mediums must:

(a) Be a liquid (c) Be non-corrosive

(b) Be a gas (d) Be in the right temperature range

214. Gas (R-12) enters the condenser at 116.9 psig 100F. It leaves the condenser at 80°F. Its pressure will be:

(a) 76.9 psig (c) 91.7 psig

(b) 84.1 psig (d) 116.9 psig

215. The capacity of an evaporative condenser depends on:

(a) Fan horsepower

(b) The amount of heat leaving air is capable of absorbing.

(c) Entering air wet bulb

(d) The temperature of the entering air

216. The Evaporator is a device to:

(a) Store liquid refrigerant

(b) Absorb heat into the refrigeration system

(c) Remove heat from the refrigeration system

(d) Keep the compressor busy

217. The temperature of the medium cooled must be:

(a) Below evaporator temperature

(b) Equal to evaporator temperature

(c) Above evaporator temperature

(d) Below freezing

218. The heat picked up in the evaporator must equal:

 (a) The heat of compression

 (b) The heat given up by the condenser

 (c) The heat lost by the medium being cooled

 (d) The heat gained by the condenser

219. Flooded feeds are different from direct expansion feeds in that they:

 (a) Are not as efficient

 (b) Use bigger evaporators

 (c) Allow for reguririgitation of liquid refrigerant

 (d) Allow for recirculation of liquid refrigerant

220. The metering device is designed to:

 (a) Keep the evaporator as cold as possible

 (b) To control evaporator superheat

 (c) Control the flow of refrigerant

 (d) To keep a constant evaporator pressure

221. The thermostatic expansion valve has three operating pressures:

 (a) Evaporator pressure, spring pressure, suction pressure

 (b) Evaporator pressure, bulb pressure, condensing pressure

 (c) Evaporator pressure, spring pressure, political pressure

 (d) Evaporator pressure, spring pressure, bulb pressure

222. The "heat of compression" is:

(a) Carried away in the evaporator

(b) Sometimes wasted

(c) Less on water-cooled compressors

(d) Carried away in the cooling medium leaving the condenser

223. The bellows type-pressure control is designed for:

(a) "lo" side use

(b) "hi" side use

(c) "lo" and "hi" side use

(d) temperature control

224. Latent heat and refrigeration effect can be the same for a refrigerant.

(a) True

(b) False

225. The pour point of refrigerant oil should never be below the evaporator temperature

(a) True

(b) False

226. In reference to refrigerant oil, viscosity is:

(a) A measure of weight

(b) The coefficient of external friction

(c) A measure of heat

(d) A way of telling how thick an oil is

227. As the percentage of R-12 in oil increases:

(a) The viscosity of the oil decreases

(b) The refrigerant starts to separate

(c) The viscosity of the mixture decreases

(d) The viscosity of the misture increases

228. An oil separator will keep all oil from circulating through the refrigerant system.

(a) True (b) False

229. An oil separator should be kept cool at all times.

(a) True (b) False

230. The material within a strainer drier is known as a:

(a) Designate (c) Desiccant

(b) Filling (d) Desecrate

231. A sight glass in a full liquid line will be:

(a) Full of bubbles (c) Light green

(b) Cloudy (d) Clear

232. An evaporator pressure regulator is designed to maintain a constant pressure of temperature in the evaporator:

(a) Regardless of how high the compressor suction pressure may go

(b) Regardless of how high the condenser pressure may go

(c) Regardless of how low the compressor suction pressure may go

(d) Regardless of how low the condenser pressure may go

233. Horizontal piping should be run level.

(a) True (b) False

234. Pressure drop is important in suction lines.

(a) True (b) False

235. The compressor compresses the liquid refrigerant:

(a) True (b) False

236. Humidistat control elements are sometimes made of:

(a) Cat-gut (c) Horse hair

(b) Bi-metal (d) Human hair

237. How much should the overall drop (pitch) be for a 50 foot steam main?

(a) 2-1/2" (c) 1"

(b) 3" (d) 5"

238. Which one of these methods of heat transfer is not employed by the various types of heat transfer units:

(a) Radiation (c) Gravitation

(b) Convection (d) Conduction

239. A material which resists the transfer of heat is called:

(a) A coil (c) A conductor

(b) A convector (d) An insulator

240. Which one of the following is not a common type of steam heating system:

(a) Vapor (c) Vacuum

(b) Atmospheric (d) One pipe gravity

241. Steam separators separate the:

(a) Water from steam

(b) Oil from steam

(c) High pressure from low pressure steam

(d) Saturated steam from superheated steam

242. Adequate sub-cooling is normally provided by condensers.

    (a) True   (b) False

243. Use a check valve in hot gas lines when the:

    (a) The compressor is in a cold location

    (b) Condenser is down cellar

    (c) Compressor is below the condenser

    (d) Pipe size is over 1-1/2"

244. The chemical moisture indicator, when encased in a refrigerant sight glass, must be filled with dry liquid refrigerant before it will function.

    (a) True   (b) False

245. When using the gauge scale to read pressures below atmospheric, the higher the number the lower the pressure.

    (a) True   (b) False

246. A wet bulb vacuum indicator records:

    (a) The temperature of the vacuum

    (b) The pressure of the vacuum

    (c) The temperature of boiling water

    (d) Condensate temperature

247. A charging manifold assembly will consist of a gauge, a manifold and three hoses.

    (a) True   (b) False

248. When charging by the sight glass method, the unit should be under full load conditions.

    (a) True   (b) False

249. Heat, in a concentrated form or at high temperature, should never be applied to a refrigerant bottle.

(a) True (b) False

250. When removing reusable refrigerant from a system, the line to the refrigerant storage drum must:

(a) Contain a sweat fitting

(b) Be flexible

(c) Be of copper

(d) Contain a strainer drier

251. When releasing a halogenated hydrocarbon refrigerant to the atmosphere, ample ventilation must be provided and,

(a) As the gas is toxic, gassmasks should be worn

(b) No open flames should be permitted

(c) Explosion precautions must be taken

(d) The owner should be informed

252. When charging through a suction manifold and compressor suction service valve, the manifold valve position should probably be as follows:

(a) Low side, front seated; high side, back seated

(b) Low side, back seated; high side, front seated

(c) Low side, front seated; high side, front seated

(d) Low side, back seated; high side, back seated

253. Consideration must always be given to the movement of air around a water cooled condenser.

(a) True (b) False

254. Any good copper tubing is satisfactory for refrigeration purposes.

(a) True (b) False

255. When installing crankcase equalizing lines, the gas equalizer may be at or above the normal oil level and the oil equalizing line may be at or below the oil level.

(a) True (b) False

256. Suction risers to individual compressors are sized for oil return under maximum load.

(a) True (b) False

257. When mufflers are used in multiple compressor installations, they should normally be mounted ______________.

(a) In the horizontal line leaving each compressor

(b) In the vertical line from each compressor to the common header

(c) In the common header downstream of the last compressor

(d) In the first vertical run of the common header

258. A diverting valve usually has:

(a) 2 inlets

(b) 2 inlets - 1 outlet

(c) 1 inlet - 2 outlets

(d) 1 inlet - 1 outlet

259. The boiling point of a refrigerant is determined by:

(a) Atmospheric pressure

(b) Outside air temperature

(c) Pressure exerted upon it

(d) Size of the compressor

260. The specific heat of water is:

(a) .5

(c) 1.0

(b) .24 (c) 1.5

261. The boiling point of a liquid increases when:

(a) The pressure is lowered

(b) The pressure is increased

(c) More heat is added

(d) Its temperature is lowered

262. When 2" pulley is making 75 rpm, what is the rpm of the 3" pulley?

(a) 100 rpm (c) 50 rpm

(b) 150 rpm (d) 125 rpm

263. On an F-12 Refrigeration System, if the suction pressure at the compressor is 35 psi with a 2 psi suction line loss and the suction line temperature taken at point TX Valve Bulb is fastened is 51°, what is your Superheat Reading?

p

(a) 13° F (c) 11° F

(b) 14° F (d) 9° F

264. A grille equipped with a damper or control valve is a:

(a) Vane (c) Damper

(b) Register (d) Louver

265. On a steam system what valve should be used in a horizontal line where drainage is required:

(a) Any valve (c) Globe valve

(b) Gate valve (d) Check valve

266. How much should the overall drop (pitch) be for a 50 foot steam main?

(a) 2-1/2" (c) 1"

(b) 3" (d) 5"

267. The lowest safe water line in a boiler should not be lower than what part of the gage glass:

(a) When no water is visible

(b) Lowest visible part

(c) Highest visible part

(d) Middle part

268. In using oxyacetylene welding equipment how should the cylinders be placed:

(a) Upright position

(b) Horizontal position

(c) Inclined position

(d) Makes absolutely no difference

269. When condensate backs up in the return lines because of lack of proper head between the dry return and boiler water level, the water line in a steam boiler will:

(a) Rise several inches

(b) Rise

(c) Not vary

(d) Drop

270. A device to protect heating plants from excessive temperature or pressure is a:

(a) High pressure control

(b) Limit control

(c) Blow down line

(d) Thermostat

271. On a steam heating system one purpose of a flash tank is to:

(a) Generate steam

(b) Add heat to the condensate

(c) Remove heat from the condensate

(d) Add heat to the steam

272. By what means is a gas burner provided with a safety device:

(a) Slenoid valve
(b) Thermo-couple
(c) Thermostatic valve
(d) Water safe-flow

273. Which of the following is *not* a type of steam trap?

(a) Bucket trap

(b) Float and thermostatic trap

(c) P-trap

(d) Impulse trap

274. Which of the following is a type of steam heating system:

(a) Direct expansion
(b) Vacuum
(c) Reverse return
(d) Injection

275. What is the suction pressure reading for 5° F for refrigerant 12:

(a) 19.5
(b) 6.5
(c) 11.8
(d) 2.8

276. On refrigeration systems all pressure relief devices (not fusible plugs) shall be:

(a) Directly pressure-actuated

(b) Remote pressure-actuated

(c) Not pressure actuated

(d) Manually operated

277. When pressure of high and low sides are same with machine operating what is apt to have happened to compressor:

   (a) Gas has leaked out

   (b) Suction and discharge valves broken

   (c) Piston rings broken

   (d) Gaskets have slipped

278. How many copper tubing be softened (annealed)?

   (a) By pounding

   (b) By bending

   (c) By heating to a blue color and cooling

   (d) By freezing

279. Which is the most brittle metal used in refrigeration system mechanisms?

   (a) Copper
   (c) Aluminum

   (b) Cast iron
   (d) Steel

280. What should be done before a gauge manifold is removed from a system?

   (a) The pressure in all parts of the manifold and its connections should be balanced

   (b) It is not necessary to balance the pressure in all parts of the manifold

   (c) It is only necessary to turn the SSV and the DSV all the way out

   (d) It is only necessary to turn the manifold valves all the way in

281. A TX Valve in a refrigeration system is controlled by a bulb located:

   (a) In the liquid line

   (b) At inlet to the evaporator

(c) On suction line as it leaves the evaporator

(d) On suction line as it enters the evaporator

282. Of what metals is silver solder usually made?

(a) Silver, lead and zinc

(b) Silver, copper and zinc

(c) Silver, copper and tin

(d) Silver, lead and tin

283. How is the motor capacitor connected to the starting winding electrically:

(a) In series

(b) In parallel

(c) Either way

(d) Between the starting and running winding

284. How is a voltmeter connected into the circuit?

(a) In series

(b) Either way

(c) In parallel

(d) Both in series and parallel

285. How must an oil separator be mounted?

(a) Level

(b) Suspend from the condenser line

(c) Below the compressor

(d) Above the compressor

286. Why do some thermostatic expansion valves (TXV) systems, in good condition, frost back when they are first started?

(a) The thermostats are slow to open

(b) There is residual oil in the cooling coil

(c) The TXV needle always leaks a little

(d) The TXV hunts or sways until it settles

287. Where does the moisture freeze in a TXV system?

(a) On the screen
(b) In the liquid line
(c) On the valve orifice
(d) In the cooling coils

288. On the high side of an ammonia (NH_2) piping system, test pressure should not be less than:

(a) 300 lbs
(b) 250 lbs
(c) 200 lbs
(d) 150 lbs

289. Circumference of pipe equals:

(a) Diameter x 3.1416
(b) π x radius square (πr^2)
(c) Radius x 3.1416
(d) Diameter x 6.2831

290. A pipe fitting shaped like an ell, but with one female end and one male end is called a:

(a) Male union ell
(b) Female union ell
(c) Street ell
(d) Female to male ell

291. When making up a pipe joint where you are putting a valve on an 8" long pipe nipple, it is considered good practice to:

(a) Put the valve in the vise and make up nipple with pipe wrench
(b) Put the pipe nipple in the vise and make up valve with pipe wrench
(c) You may either put valve or nipple in vise and make up other with pipe wrench
(d) Don't use vise to make up this kind of joint, use 2 pipe wrenches

292. Find the length of a piece of pipe for a 90° bend with a radius of 40 in. and with two 15 in. tangents.

(a) 65.82 in. (c) 77.82 in.

(b) 92.82 in. (d) 62.82 in.

293. In designating the outlets of reduced fittings, whether of the flanged or screwed type, the openings should be read in a certain order. Therefore, on side outlet reducing fitting the size of the side outlet is named in what order:

(a) First (c) First or last

(b) According to size (d) Last

294. A double offset expansion bend is shown in Figure 1.8. If R = 21 in. H = 26', Right tan = 1'6" Left tan = 1'2", then the developed length of pipe is ________.

(a) 198 in. (c) 242 in.

(b) 230 in. (d) 21-1/2 ft

295. In Figure 2.6, if W = 24-3/4 in., R = 30 in., Right tan = 30", Left tan = 18", then the developed length of pipe is ________.

(a) 28-1/4 ft (c) 331 in.

(b) 283 in. (d) 24-3/4 ft

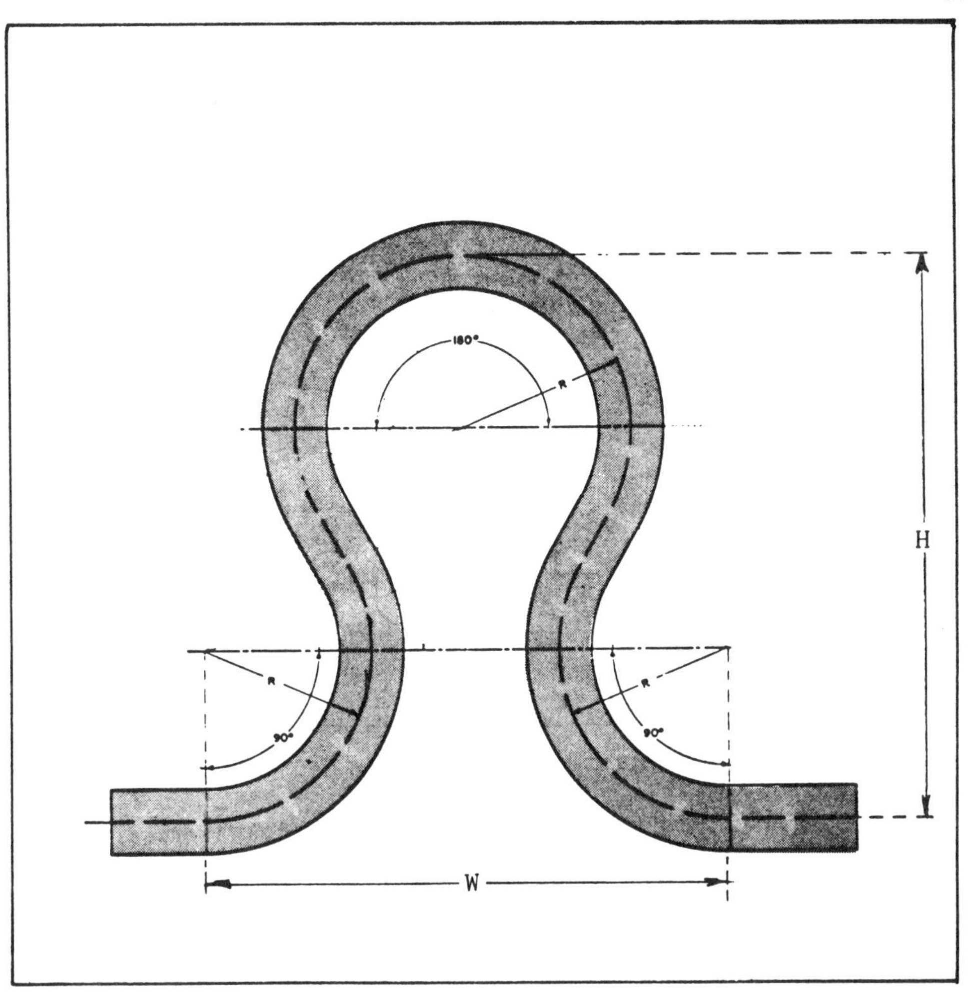

FIGURE 1.8

In the next 5 questions, identify the types of cooling systems shown in Figure 1.9 as follows;

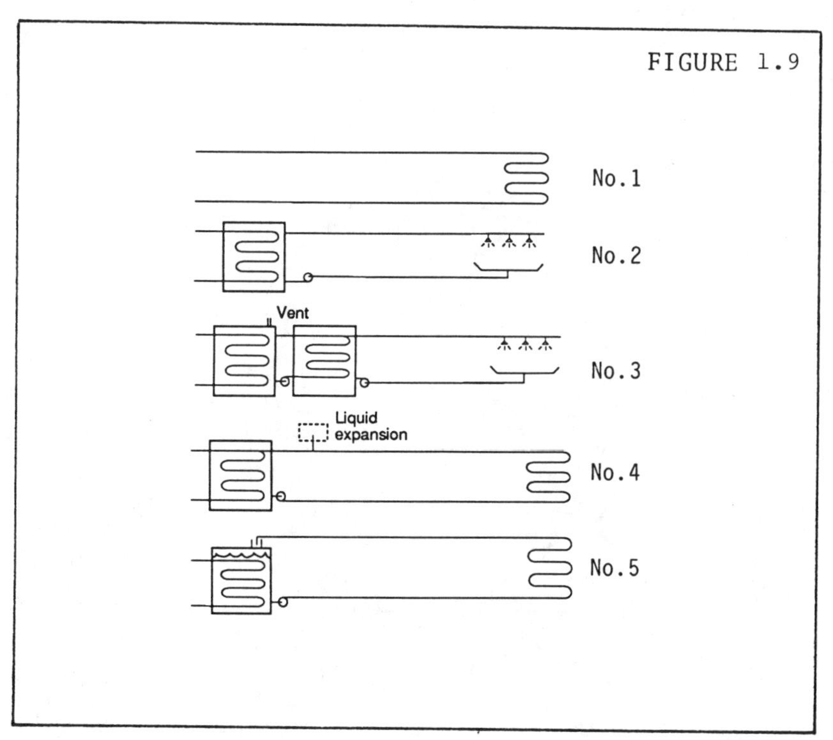

296. No. 1 is:

 (a) Direct system

 (b) Indirect system

 (c) Indirect vented system

 (d) Brine system

297. No. 2 is:

 (a) Direct expansion system

 (b) Double indirect vented system

 (c) Indirect vented closed surface system

 (d) Indirect open spray system

298. No. 3 is:

(a) Direct closed surface system

(b) Indirect closed surface system

(c) Double indirect vented open spray system

(d) Indirect open spray system

299. No. 4 is:

(a) Direct closed surface system

(b) Indirect closed surface system

(c) Indirect vented closed surface system

(d) Double indirect vented open spray system

300. No. 5 is:

(a) Direct closed surface system

(b) Indirect closed surface system

(c) Double indirect vented open spray system

(d) Indirect vented closed surface system

PART 2

TECHNICAL TABLES & FORMULAS

BASIC MATHEMATICS

FRACTIONS

To Reduce Common Fractions: divide the numerator and denominator by common divisors until further reduction is impossible -

$$63/81 = \frac{21}{3\overline{)63}} \quad \frac{27}{3\overline{)81}} = 21/27 = 7/9$$

To Reduce Improper Fractions: divide the numerator by the denominator, the quotient being a whole number and and the remainder the new numerator -

$$43/6 = 6\overline{)43} = 7\text{-}1/6$$

To Express a Fraction as a Decimal: divide the numerator by the denominator -

$$3/4 = 4\overline{)3} = \frac{.75}{4\overline{)3.00}} = 0.75$$

To Reduce Complex Fractions: first express both numerator and denominator as simple fractions, then multiply the upper numerator by the lower denominator for the new numerator and the lower numerator by the upper denominator -

$$\frac{1\text{-}3/4}{5/6} = \frac{7/4}{5/6} = 42/20 = 20\overline{)42} = 2\text{-}1/10$$

To Reduce Fractions to a Common Denominator: multiply the numerator of each fraction by the product of all of the denominators except its own for the new numerators and multiply all denominators together for the new common denominator -

$$2/3,\ 1/4,\ 3/5 = 40/60,\ 15/60,\ 36/60$$

To Add Fractions: reduce to a common denominator and add the numerators -

$$3/4 + 2/3 = 9/12 + 8/12 = 17/12 = 1\text{-}5/12$$

To Subtract Fractions: reduce to a common denominator and subtract the numerators -

3/4 - 2/3 = 9/12 - 8/12 = 1/12

To Multiply Fractions: multiply the numerators for a new numerator and multiply the denominators for a new denominator -

3/4 x 5/8 = 15/32

To Divide Fractions: invert the divisor and multiply

a. 3/4 ÷ 7/8 = 3/4 x 8/7 = 24/28 = 6/7

b. 1-1/8 ÷ 3/16 = 9/8 ÷ 3/16 = 9/8 x 16/3 = 144/24 = 6/1 = 6

DECIMALS

To Express a Decimal as a Fraction: disregard the decimal point and write the figures as the numerator of the fraction. Write the denominator as 1 plus as many zeros after it as there were figures following the decimal -

$$.0125 = \frac{125}{10000}$$

To Express a Fraction as a Decimal: divide the numerator by the denominator -

$$32/64 = 1/2 = 2\overline{)1.0}^{\;.5} = .5$$

To Multiply Decimals: proceed as in simple multiplication and point off, as many decimal places in the product as are in the multiplier and multiplicand together -

```
$87.96 x 23.5 =      87.96   =  3 places
                   x 23.5
                 ---------
                     43980
                    26388
                   17592
                 ---------
                 $ 2067.060 ( 3 places )
```

To Divide Decimals: proceed as in simple division and point off, as many decimal places in the quotient as are in the dividend in excess of the divisor -

```
                                .67
0.2546 ÷ 0.38 =  .38 / .25 46  =  0.67
                        22 8
                        26 6
                        26 6
```

RATIO AND PROPORTION

Ratio is the relation of one figure to another and is sometimes expressed as a fraction with the first quantity as the numerator:

The ratio of 1 to 2 = 1:2 = 1/2

When ratios are equal to each other they are said to be in proportion. The ratio of 3 to 6 = 3:6 = 1/2, therefore it is equal to and in proportion to the ratio of 1 to 2 and the proportion would be written,

$$3:6 = 1:2$$

and, read, " 3 is to 6 as 1 is to 2 ".

The first and last terms in a statement of proportion are called the *extremes*, and the middle terms are the *means*. A rule of proportion is that the product of the extremes is equal to the product of the means; therefore, in the above example;

3:6 = 1:2 = 3 x 2 = 6 and 6 x 1 = 6

When the middle terms are identical this quantity is called the mean proportional of the first and last terms;

1:2 = 2:4, 2 is the mean proportional between 1 and 4. To find the mean proportional of any two terms multiply them and extract the square root of their product. Thus the mean proportional of 2 and 50 is;

$\sqrt{2 \times 50} = \sqrt{100} = 10$. Therefore, 2:10 = 10:50

If proportion is expressed algebraically as;

$$a:b = c:d$$

then given any three terms the fourth can be determined. In direct proportions we may "solve for χ" by one of the following formulas:

1.) $a = \frac{b\ c}{d}$ 3.) $b = \frac{a\ d}{c}$

2.) $c = \frac{a\ d}{b}$ 4.) $d = \frac{b\ c}{a}$

In the following examples, solve for χ by using the above formulas;

1. 24:4 as χ:3 = $c = \frac{a\ d}{b}$

$= \frac{24x3}{4} = \frac{72}{4} = 18 =$ 24:4 as 18:3

2. 2:3 as 4:χ = $d = \frac{b\ c}{a}$

$= \frac{3x4}{2} = \frac{12}{2} = 6 =$ 2:3 as 4:6

3. 5:χ 25:20 = $b = \frac{a\ d}{c}$

$= \frac{5x20}{25} = \frac{100}{25} = 4 =$ 5:4 as 25:20

4. χ:25 as 10:2 = $a = \frac{b\ c}{d}$

$= \frac{25x10}{2} = \frac{250}{2} = 125 =$ 125:25 as 10:2

Direct proportion formulas are useful in solving many problems, for example;

5. A 20 ft wall, shades 3/4 of a lawn, how much higher must the wall be to shade the entire lawn?

20:3/4 as χ:1 = 20:0.75 as χ:1 =

$\frac{20x1}{.75} =$ 26.6 26.6 minus 20 = 6.6 ft

SQUARE AND SQUARE ROOTS

To Find the Square of any Number: multiply the number by itself -

The superscript^2 denotes *square*, thus 4 squared is written;

$$4^2 = 16$$

, and 256 squared is written;

$$256^2 = 65536$$

To Find the Square Root of any Number: by trial and error, determine which number when multiplied by itself will equal the number given -

The symbol $\sqrt{\quad}$ denotes square root, thus the square root of 16 is written;

$$\sqrt{16} = 4$$

, and the square root of 2302.625 is written;

$$\sqrt{2302.625} = 47.98$$

Finding the square root of numbers of 3 digits or more is a slow and laborious task, the examination candidate should either use a slide rule or refer to a standard table. Table 3.1 gives the square and square root for numbers from 1 through 100.

TABLE 2.1

SQUARES AND SQUARE ROOTS

No.	Square	Square Root	No.	Square	Square Root	No.	Square	Square Root	No.	Square	Square Root
1	1	1.0000	26	676	5.0990	51	2601	7.1414	76	5776	8.7178
2	4	1.4142	27	729	5.1962	52	2704	7.2111	77	5929	8.7750
3	9	1.7321	28	784	5.2915	53	2809	7.8201	78	6084	8.8318
4	16	2.0000	29	841	5.3852	54	2916	7.3485	79	6241	8.8882
5	25	2.2361	30	900	5.4772	55	3025	7.4162	80	6400	8.9443
6	36	2.4495	31	961	5.5678	56	3136	7.4833	81	6561	9.0000
7	49	2.6458	32	1024	5.6569	57	3249	7.5498	82	6724	9.0554
8	64	2.8284	33	1089	5.7446	58	3364	7.6158	83	6889	9.1104
9	81	3.0000	34	1156	5.8310	59	3481	7.6811	84	7056	9.1652
10	100	3.1623	35	1225	5.9161	60	3600	7.7460	85	7225	9.2195
11	121	3.3166	36	1296	6.0000	61	3721	7.8102	86	7396	9.2736
12	144	3.4641	37	1369	6.0828	62	3844	7.8740	87	7569	9.3276
13	169	3.6056	38	1444	6.1644	63	3969	7.9373	88	7744	9.3808
14	196	3.7417	39	1521	6.2450	64	4096	8.0000	89	7921	9.4340
15	225	3.8730	40	1600	6.3246	65	4225	8.0623	90	8100	9.4868
16	256	4.0000	41	1681	6.4031	66	4356	8.1240	91	8281	9.5394
17	289	4.1231	42	1764	6.4807	67	4489	8.1854	92	8464	9.5917
18	324	4.2426	43	1849	6.5574	68	4624	8.2462	93	8649	9.6437
19	361	4.3589	44	1936	6.6332	69	4761	8.3066	94	8836	9.6954
20	400	4.4721	45	2025	6.7082	70	4900	8.3666	95	9025	9.7468
21	441	4.5826	46	2116	6.7823	71	5041	8.4261	96	9216	9.7980
22	484	4.6904	47	2209	6.8557	72	5184	8.4853	97	9409	9.8489
23	529	4.7958	48	2304	6.9282	73	5329	8.5440	98	9604	9.8995
24	576	4.8990	49	2401	7.0000	74	5476	8.6023	99	9801	9.9499
25	625	5.0000	50	2500	7.0711	75	5625	8.6603	100	10000	10.0000

SQUARE ROOT OF FRACTIONS	
Fraction	Square Root
1/8	.3535
1/4	.5000
3/8	.6124
1/2	.7071
5/8	.7906
3/4	.8660
7/8	.9354

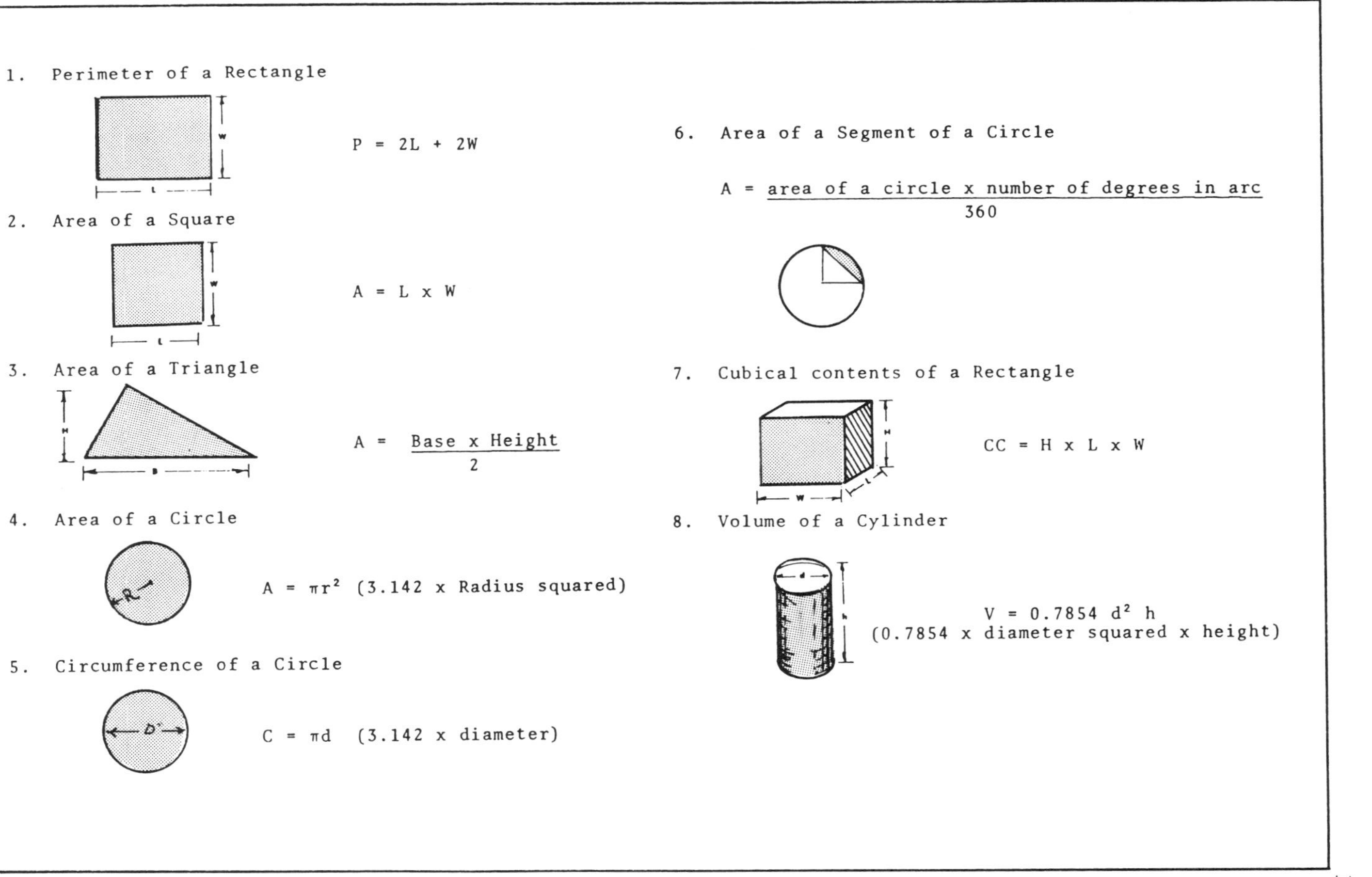

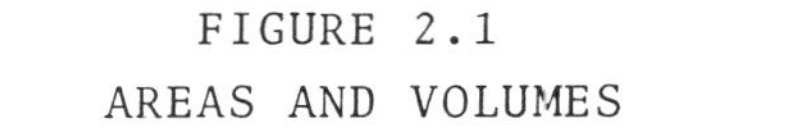

FIGURE 2.1
AREAS AND VOLUMES

TABLE 2.2

IMPORTANT THINGS TO REMEMBER

One gallon of water = 8.34 pounds

One pound of water = 7000 grains

One cubic ft H_2O = 62.37 pounds

One psi = 2.31 feet of water

One ft of water = 0.433 psi

Atmospheric pressure = 14.7 psi

Absolute pressure (psia) = psig + 14.7

Gauge pressure (psi) = psia - 14.7

One watt = 3.415 Btuh

One kilowatt = 3415 Btuh

Absolute temperature = °F + 460 : Absolute K = -460F

Fahrenheit (F) = (9/5 x C) + 32 or,
(Celsius x 1.8) + 32 = Fahrenheit

Celsius(C) = (F - 32) x 5/9 or,
(Fahrenheit - 32) x .555 = Celsius

Refrigerant 12 = Dichlorodifluoromethane = CCl_2F_2

Refrigerant 500 = Dichlorodifluoromethane 73.8% = CCl_2F_2/CH_3 CHF_2

Refrigerant 22 = Chlorodifluoromethane = $CHClF_2$

The above list of IMPORTANT THINGS TO REMEMBER must be committed to memory. This is the basic remember list for any candidate for the Journeyman License Examination.

IMPORTANT FORMULAE FOR SOLVING BASIC PROBLEMS

AIR FLOW

When solving air flow problems for duct systems, the calculation for velocity, cfm and duct sizing is best accomplished by the use of the friction chart or Ductulator. But a duct system is always figured on the ratio of its length to 100 equivalent ft.

To solve air flow problems at any given point in a system or under any other condition, use the following formulas;

1. cfm = fpm x area, ft^2
2. fpm = cfm/area, ft^2
3. area = cfm/fpm

This may be easily remembered by the air circle equation

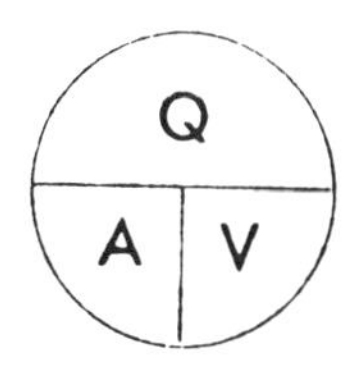

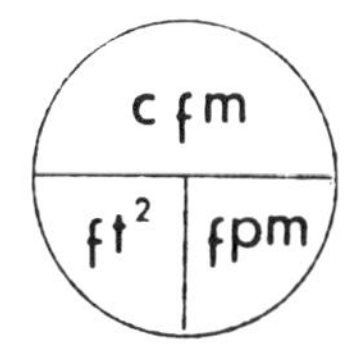

To find the quantity (Q) of air, the velocity (V) of air, or the area (A) through which the air passes, cover that which is to be found--with your thumb--and follow the signs.

AIR CHANGES

1. To find the quantity of outside air where the number of outdoor air changes per hour are specified;

$$\text{cfm} = \frac{\text{space, ft}^3 \text{ x number of changes}}{60 \text{ min}}$$

2. To find the number of air changes per hour where the air quantity, cfm, is known;

$$\text{number of air changes/hr} = \frac{\text{cfm x 60 min}}{\text{space, ft}^3}$$

3. To find the air quantity, cfm, required per air changes;

$$\text{cfm per air change} = \frac{\text{space, ft}^3}{\text{minutes per change}}$$

FORMULAS FOR CONVERTING PRESSURE

When pressure is read on a barometer it is called absolute pres- sure; pressure read on a gauge is called atmospheric pressure. Atmospheric pressure will read 14.7 psia on a barometer.

1 atmosphere =
- 29.29 in. mercury, Hg
- 14.7 lb/sq in., psia
- 34 ft, water, 'WG
- 407 in., water, "WG

That is to say, the pressure of the atmosphere, 14.7 psia, coul support a column of water 34 feet high. The accompanying figur shows a Bourdon-type compound gauge. The scale reads zero at one atmosphere. Below one atmosphere, it reads inches of mercu vaccum—0 to 30. Above one atmosphere, it reads pounds per square inch gauge, psig.

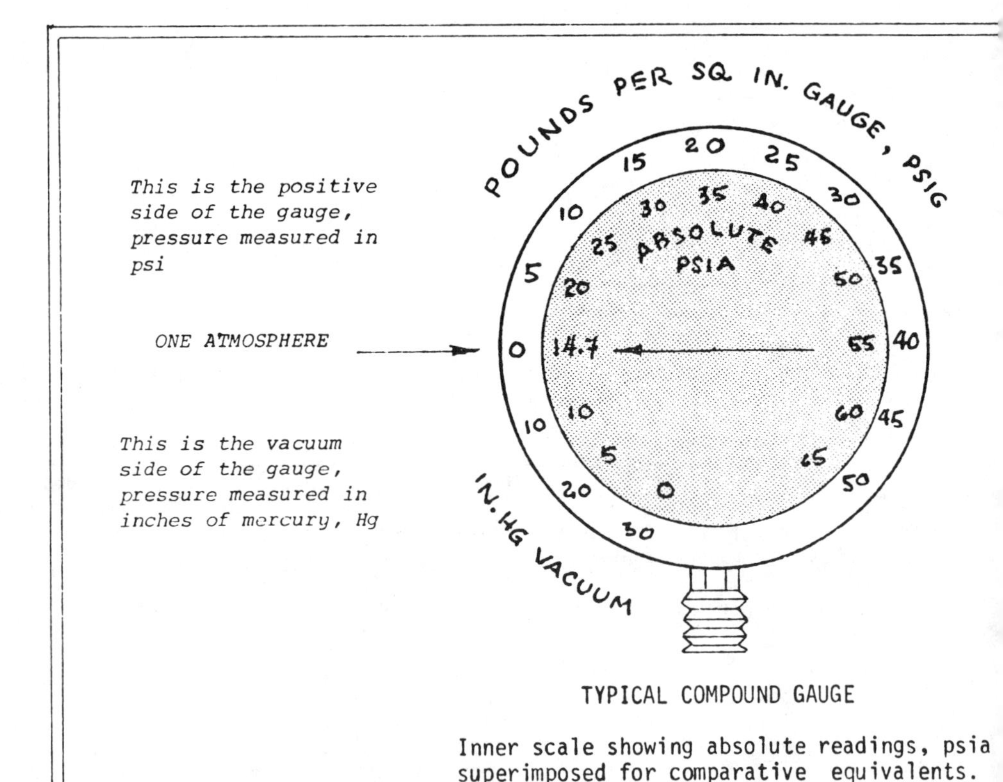

FIGURE 2.2

TABLE 2.3

FORMULAS FOR SOLVING REFRIGERATION PROBLEMS

		Equation No.
Net Refrigerating Effect, Btu/lb	$= \text{Enthalpy of Vapor Leaving Evaporator, Btu/lb} - \text{Enthalpy of Liquid Entering Evaporator, Btu/lb}$	*9.1*
Compression Work, Btu/min	$= \text{Heat of Compression, Btu/lb} \times \text{Refrigerant Circulated, lb/min}$	*9.2*
Compression Horsepower	$= \dfrac{\text{Compression Work, Btu/min}}{42.4}$	*9.3*
Compression Horsepower	$= \dfrac{\text{Capacity, Btu/min}}{42.4 \times \text{COP}}$	*9.4*
Compression Horsepower per Ton	$= \dfrac{4.715}{\text{Coefficient of Performance}}$	*9.5*
Power, watts	$= \text{Compression Horsepower per Ton} \times 745.7$	*9.6*
Coefficient of Performance	$= \dfrac{\text{Net Refrigerating Effect, Btu/lb}}{\text{Heat of Compression, Btu/lb}}$	*9.7*
Capacity, Btu/min	$= \text{Refrigerant Circulated, lb/min} \times \text{Net Refrigerating Effect, Btu/lb}$	*9.8*
Compressor Displacement, ft³/min	$= \dfrac{\text{Capacity, Btu/min} \times \text{Volume of Gas Entering Compressor, ft}^3\text{/lb}}{\text{Net Refrigerating Effect, Btu/lb}}$	*9.9*
Heat of Compression, Btu/lb	$= \text{Enthalpy of Vapor Leaving Compressor, Btu/lb} - \text{Enthalpy of Vapor Entering Compressor, Btu/lb}$	*9.10*
Volumetric Efficiency	$= 100 \times \dfrac{\text{Actual Weight of Refrigerant}}{\text{Theoretical Weight of Refrigerant}}$	*9.11*
Compression Ratio	$= \dfrac{\text{Head Pressure, psia (absolute)}}{\text{Suction Pressure, psia (absolute)}}$	*9.12*
Refrigerant Circulated, lb (min) (ton)	$= \dfrac{200}{\text{Refrigerating Effect}}$	*9.13*

42.4 = heat flow, Btu/(min) (hp); 200 = Btu/(min) (ton); COP = coefficient of performance

FORMULAE

The following symbols are common usage in electrical formulas:

I	=	Ampere, a unit of current
E	=	Volt, a unit of pressure
W	=	Watt, a measure of power
R	=	Ohm, a unit of resistance
eff	=	Efficiency; use 0.85 when eff is unknown
pf	=	Power factor; the ratio of the actual power to the apparent power
kw	=	1000 watts, a measure of power
hp	=	horsepower; 1 hp = 0.746 kw

Figure 2.3 shows the Power Equation Wheel from which the basic electrical formulas may be read. W may be substituted for P (power). Symbols in the outer ring equal the symbol in the inner ring.

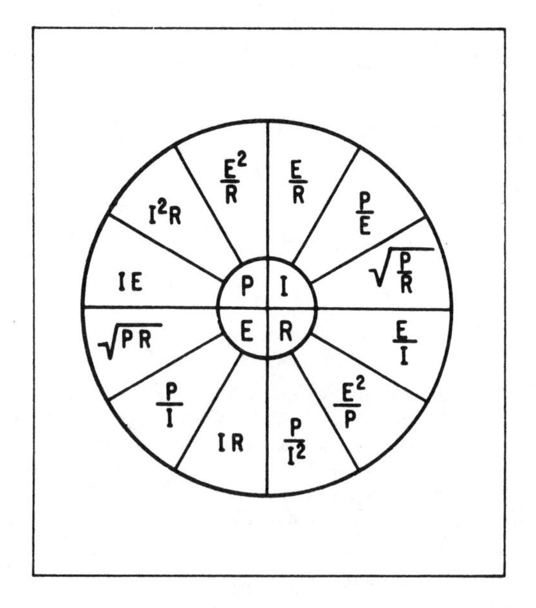

FIGURE 2.3

To Find the Power Factor:

. Single phase circuits; $pf = \frac{W}{E \times I}$

. Two phase circuits; $pf = \frac{W}{E \times I \times 2}$

. Three phase circuits; $pf = \frac{W}{E \times I \times 1.73}$

To Find Amps Where Hp is Known:

. Single phase circuits; $I = \frac{hp \times 746}{E \times eff \times pf}$

. Three phase circuits; $I = \frac{hp \times 746}{1.73 \times E \times eff \times pf}$

To Find Hp where Amps are Known:

. Single phase circuit; $hp = \frac{I \times E \times eff \times pf}{746}$

. Three phase circuit; $hp = \frac{I \times E \times eff \times pf \times 1.73}{746}$

IREE PHASE AC CIRCUITS

Three phase is the most common polyphase system. It is onnected in either "delta" or "star" (sometimes called Y) ormation. Figure 2.4 shows a *delta* system, the current etween any two wires is 240 V; the line voltage in a *delta* onnection equals the *coil voltage*. Figure *4.8* shows a *tar* system, the current between the neutral and any hot ire is single-phase 120 volts; the line voltage in a *star* onnection equals the *coil current*.

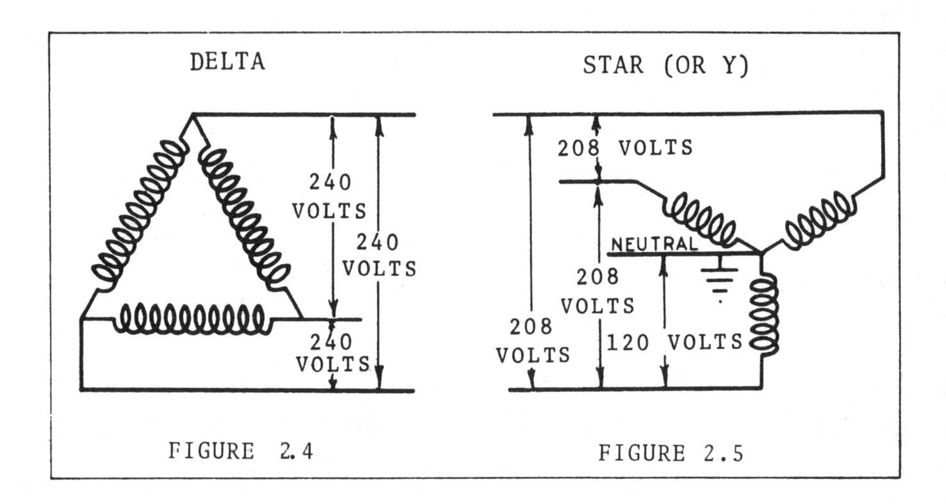

FIGURE 2.4

FIGURE 2.5

ELECTRICAL MEASUREMENTS

Electrical units of measurement are given in Table 2.4 together with the instruments generally used to measure each unit. Frequency, inductance (henry), and capacitance (farad) may also be measured but will not be discussed here.

TABLE 2.4

PRACTICAL ABSOLUTE UNITS OF ELECTRICAL QUANTITIES

CODE	DESCRIPTION	SYMBOL	UNIT	MEASURING INSTRUMENT
Voltage (emf)	Pressure	V	Volt	Voltmeter
Resistance		Ω	Ohm	Ohmeter
Current	Flow	I	Ampere	Ammeter
Power	Rate of Energy	W	Watt	Wattmeter
Power Factor	Power Ratio	pf		Power Factor Meter

In Figure 2.6 the application for measuring instruments is shown graphically in the circuit.

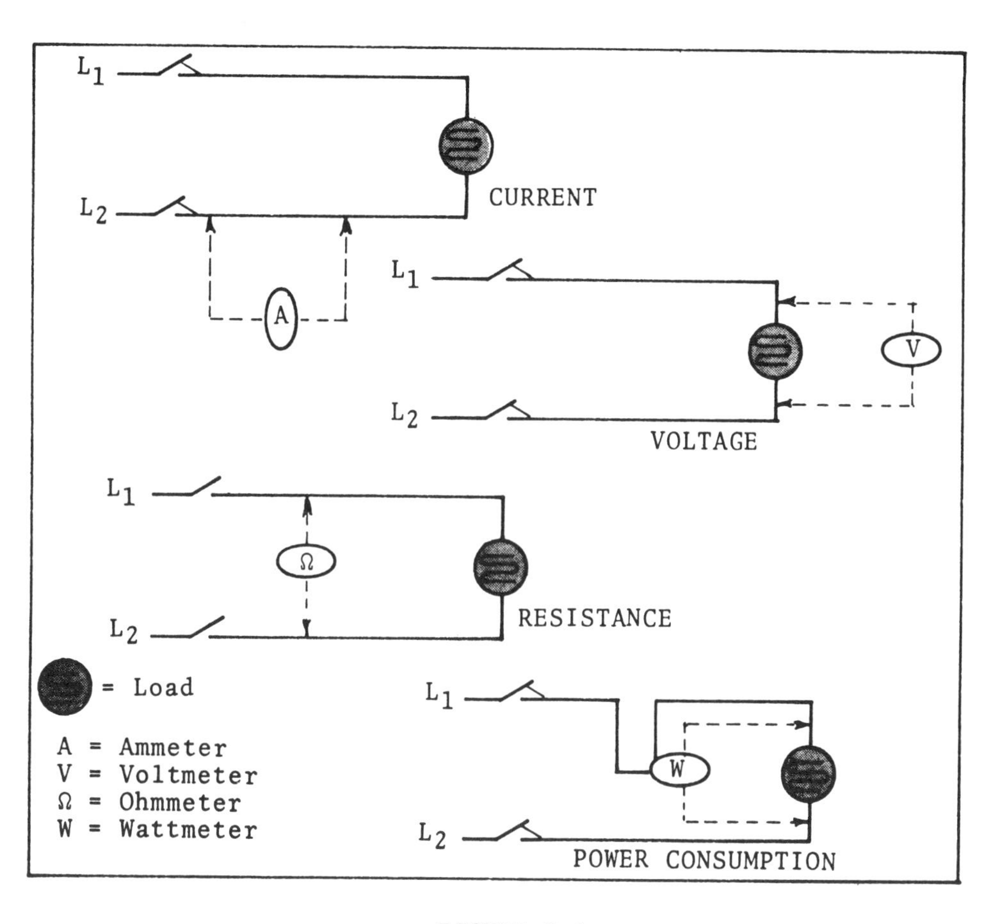

FIGURE 2.6

As a matter of convenience, a prefix system is used to express very large or very small quantities of electrical units, thereby avoiding the use of many zeroes and decimals. Some common prefixes and their equivalents are:

Kilo	=	Thousands	1 kilowatt	=	1000 watts
Meg	=	Millions	1 megohm	=	1,000,000 ohms
Mili	=	1/1000	1 miliamp	=	0.001 amp
Micro	=	1/1,000,000	1 microamp	=	0.000001 amp

TABLE 2.5

TEMPERATURE CONVERSIONS

Use this table to convert listed temperatures from either Fahrenheit to Celsius, or Celsius to Fahrenheit. For temperatures not listed, use the following formulas;

Fahrenheit = (C x 1.8) + 32 = F   Celsius = (F -32) x .555 = C

C	F	C	F	C	F	C	F	C	F
-30	-22.0	20	68.0	70	158.0	120	248.0	170	338.
-29	-20.2	21	69.8	71	159.8	121	249.8	171	339.
-28	-18.4	22	71.6	72	161.6	122	251.6	172	341.
-27	-16.6	23	73.4	73	163.4	123	253.4	173	343.
-26	-14.8	24	75.2	74	165.2	124	255.2	174	345
-25	-13.0	25	77.0	75	167.0	125	257.0	175	347
-24	-11.2	26	78.8	76	168.8	126	258.8	176	348
-23	-9.4	27	80.6	77	170.6	127	260.6	177	350
-22	-7.6	28	82.4	78	172.4	128	262.4	178	352
-21	-5.8	29	84.2	79	174.2	129	264.2	179	354
-20	-4.0	30	86.0	80	176.0	130	266.0	180	356
-19	-2.2	31	87.8	81	177.8	131	267.8	181	357
-18	-0.4	32	89.6	82	179.6	132	269.6	182	359
-17	+1.4	33	91.4	83	181.4	133	271.4	183	361
-16	3.2	34	93.2	84	183.2	134	273.2	184	363
-15	5.0	35	95.0	85	185.0	135	275.0	185	365
-14	6.8	36	96.8	86	186.8	136	276.8	186	366
-13	8.6	37	98.6	87	188.6	137	278.6	187	368
-12	10.4	38	100.4	88	190.4	138	280.4	188	37
-11	12.2	39	102.2	89	192.2	139	282.2	189	37
-10	14.0	40	104.0	90	194.0	140	284.0	190	37
-9	15.8	41	105.8	91	195.8	141	285.8	191	37
-8	17.6	42	107.6	92	197.6	142	287.6	192	37
-7	19.4	43	109.4	93	199.4	143	289.4	193	37
-6	21.2	44	111.2	94	201.2	144	291.2	194	38
-5	23.0	45	113.0	95	203.0	145	293.0	195	38
-4	24.8	46	114.8	96	204.8	146	294.8	196	38
-3	26.6	47	116.6	97	206.6	147	296.6	197	38
-2	28.4	48	118.4	98	208.4	148	298.4	198	38
-1	30.2	49	120.2	99	210.2	149	300.2	199	39
0	32.0	50	122.0	100	212.0	150	302.0	200	39
1	33.8	51	123.8	101	213.8	151	303.8	201	39
2	35.6	52	125.6	102	215.6	152	305.6	202	39
3	37.4	53	127.4	103	217.4	153	307.4	203	39
4	39.2	54	129.2	104	219.2	154	309.2	204	39
5	41.0	55	131.0	105	221.0	155	311.0	205	40
6	42.8	56	132.8	106	222.8	156	312.8	206	40
7	44.6	57	134.6	107	224.6	157	314.6	207	40
8	46.4	58	136.4	108	226.4	158	316.4	208	40
9	48.2	59	138.2	109	228.2	159	318.2	209	40
10	50.0	60	140.0	110	230.0	160	320.0	210	41
11	51.8	61	141.8	111	231.8	161	321.8	211	41
12	53.6	62	143.6	112	233.6	162	323.6	212	41
13	55.4	63	145.4	113	235.4	163	325.4	213	41
14	57.2	64	147.2	114	237.2	164	327.2	214	41
15	59.0	65	149.0	115	239.0	165	329.0	215	41
16	60.8	66	150.8	116	240.8	166	330.8	216	42
17	62.6	67	152.6	117	242.6	167	332.6	217	42
18	64.4	68	154.4	118	244.4	168	334.4	218	42
19	66.2	69	156.2	119	246.2	169	336.2	219	42

EXAMPLE: C = 6, solve for F; (6 x 1.8) + 32 = 10.8 + 32 = 42.8

TABLE 2.6

SATURATED REFRIGERANT
Temperature - Pressure Chart

Italicized figures are inches of mercury;
bold type is gauge pressure in lbs per sq in.

Temp. °F	R-717 (Ammonia)	R-11	R-12	R-22	R-500	R-502
	psig	psig	psig	psig	psig	psig
-70	*21.9*	*29.5*	*21.8*	*16.6*	*20.3*	*12.6*
-65	*20.4*	*29.3*	*20.5*	*14.4*	*18.8*	*10.0*
-60	*18.6*	*29.2*	*19.0*	*12.0*	*17.0*	*7.0*
-55	*16.6*	*29.0*	*17.3*	*9.2*	*15.0*	*3.6*
-50	*14.3*	*28.9*	*15.4*	*6.2*	*12.8*	0.0
-45	*11.7*	*28.7*	*13.3*	*2.7*	*10.4*	2.1
-40	*8.7*	*28.4*	*10.9*	0.5	*7.6*	4.3
-35	*5.4*	*28.1*	*8.3*	2.6	*4.6*	6.7
-30	*1.6*	*27.8*	*5.5*	4.8	*1.2*	9.4
-25	1.3	*27.4*	*2.3*	7.4	1.2	12.3
-20	3.6	*27.0*	0.6	10.2	3.2	15.5
-15	6.2	*26.5*	2.5	13.2	5.4	19.0
-10	9.0	*26.0*	4.5	16.5	7.8	22.8
-5	12.2	*25.4*	6.7	20.1	10.4	26.7
0	15.7	*24.7*	9.2	24.0	13.3	31.2
5	19.6	*23.9*	11.8	28.2	16.4	36.0
10	23.8	*23.1*	14.6	32.8	19.7	41.1
15	28.4	*22.1*	17.7	37.7	23.4	46.6
20	33.5	*21.1*	21.0	43.0	27.3	52.5
25	39.0	*19.9*	24.6	48.8	31.5	58.7
30	45.0	*18.6*	28.5	54.9	36.0	65.4
35	51.6	*17.2*	32.6	61.5	40.9	72.6
40	58.6	*15.6*	37.0	68.5	46.1	80.2
45	66.3	*13.9*	41.6	76.0	51.7	88.3
50	74.5	*12.0*	46.7	84.0	57.6	96.9
55	83.4	*10.0*	52.0	92.6	63.9	106.0
60	92.9	*7.8*	57.7	101.6	70.6	115.6
65	103.1	*5.4*	63.8	111.2	77.8	125.8
70	114.1	*2.8*	70.2	121.4	85.4	136.6
75	125.8	0.0	77.0	132.3	93.5	148.0
80	138.3	1.5	84.2	143.6	102	159.9
85	151.7	3.2	91.6	155.7	111	172.6
90	165.9	4.9	99.8	168.4	121	185.8
95	181.1	6.8	108.3	181.8	131	199.8
100	197.2	8.8	117.2	195.9	141	214.4
105	214.2	10.9	126.6	210.8	153	229.8
110	232.3	13.2	136.4	226.4	164	245.8
115	251.5	15.6	146.8	242.7	176	262.7
120	271.7	18.3	157.6	259.9	189	280.3

GRAPHICAL SYMBOLS FOR PLAN READING

Examinations for air conditioning journeyman usually require recognition tests for graphical symbols. These will be match-type questions where five or ten symbols will be shown with scrambled answers. Open book exams are simple; the following pages show graphical symbols in general use. In closed book exams, where the candidate must rely on memory, heavy studying is required. First try to organize your recognition response by "family." Concentrate on Valves, Piping Specialties, Air Moving Devices, Grilles and Refrigeration. Notice that symbols have a certain graphic logic which, after adequate study, often simplifies the recognition. Fire Sprinklers, Heat Exchangers and Special Duty are rarely required.

Piping

Heating

High Pressure Steam*	——HPS——
Medium Pressure Steam	——MPS——
Low Pressure Steam	——LPS——
High Pressure Return	——HPR——
Medium Pressure Return	——MPR——
Low Pressure Return	——LPR——
Boiler Blow Down	——BBD——
Condensate Pump Discharge	——CP——
Vacuum Pump Discharge	——VPD——
Makeup Water	——MU——
Air Relief Line (Vent)	——V——
Fuel Oil Flow	——FOF——
Fuel Oil Return	——FOR——
Fuel Oil Tank Vent	——FOV——
Low Temperature Hot Water Supply*	——HWS——
Medium Temperature Hot Water Supply	——MTWS——
High Temperature Hot Water Supply	——HTWS——
Low Temperature Hot Water Return	——HWR——
Medium Temperature Hot Water Return	——MTWR——
High Temperature Hot Water Return	——HTWR——
Compressed Air	——A——
Vacuum (Air)	——VAC——
Existing Piping	——(NAME)E——
Pipe to be Removed	—×—×—(NAME)—×—×—

Air Conditioning and Refrigeration

Refrigerant Discharge	——RD——
Refrigerant Suction	——RS——
Brine Supply	——B——
Brine Return	——BR——
Condenser Water Supply	——C——
Condenser Water Return	——CR——
Chilled Water Supply	——CHWS——
Chilled Water Return	——CHWR——
Fill Line	——FILL——
Humidification Line	——H——
Drain	——D——

Plumbing

Soil, Waste, or Leader (Above Grade)	————————
Soil, Waste, or Leader (Below Grade)	— — — — —
Vent	- - - - - - - - - -
Cold Water	—·—·—·—
Hot Water	—··—··—··
Hot Water Return	—···—···—···
Gas	—G————G—
Acid Waste	——ACID——
Drinking Water Flow	——DW——
Drinking Water Return	——DWR——
Vacuum (Air)	——VAC——
Compressed Air	——A——
Chemical Supply Pipes†	——(NAME†)——

Firesafety Devices

Sprinkler Heads

Upright

Pendant

Flush Mounted

Sidewall

Sprinkler Piping

Main Supplies	——S——
Control Valve	
Drain	——D——
Riser and Branch (give size)	4 in.
Alarm Check Valve	

Dry Pipe Valve

Pipe Hangers

Signal Initiating Detectors

Heat (Thermal)

Smoke

Gas

Flame

Control Panel FCP

Valves for Selective Actuators

Air Line

Ball

Butterfly

Diaphragm

Gate

Gate, Angle

Globe

Globe, Angle

Globe, Stop Check

Plug Valve

Three Way

Valve Actuators

Manual

Non Rising Stem

Outside Stem & Yoke

Lever

Gear G

Electric

Motor M

Solenoid S

Pneumatic

Motor A

Diaphragm A

Float

Hydraulic Piston H

Valves, Special Duty

Check, Swing Gate

Check, Spring S

Control, Electric-Pneumatic

Control, Pneumatic-Electric

Hose End Drain

Lock Shield

Needle

Pressure Reducing, Self-Contained

Pressure Reducing, External Pressure

Pressure Reducing, Differential Pressure

Quick Opening

Quick Closing, Fusible Link

Relief (R) or Safety (S)

S

Solenoid

S

Square Head Cock

Unclassified (number and specify)[1]

V-1

Fittings

The following fittings are shown with screwed connections. The symbol for the body of a fitting is the same for all types of connections, unless otherwise specified. The types of connections are often specified for a range of pipe sizes, but are shown with the fitting symbol where required. For example, an elbow would be:

Flanged

Screwed

Belt & Spigot

Welded[2]

Soldered

Solvent Cement

Fitting	*Symbol*
Bushing	
Cap	
Connection, Bottom	
Connection, Top	
Coupling (Joint)	
Cross	
Elbow, 90°	
Elbow, 45°	
Elbow, turned up	
Elbow, turned down	
Elbow, reducing, show sizes	2 4
Elbow, base	
Elbow, long radius	
Elbow, double branch	
Elbow, side outlet, outlet up	
Elbow, side outlet, outlet down	
Lateral	
Reducer, concentric	
Reducer, eccentric straight invert	
Reducer, eccentric straight crown	
Tee	
Tee, outlet up	
Tee, outlet down	
Tee, reducing, show sizes	2 6 4
Tee, side outlet, outlet up	
Tee, side outlet, outlet down	
Tee, single sweep	
Union	

Piping Specialties

Air Eliminator

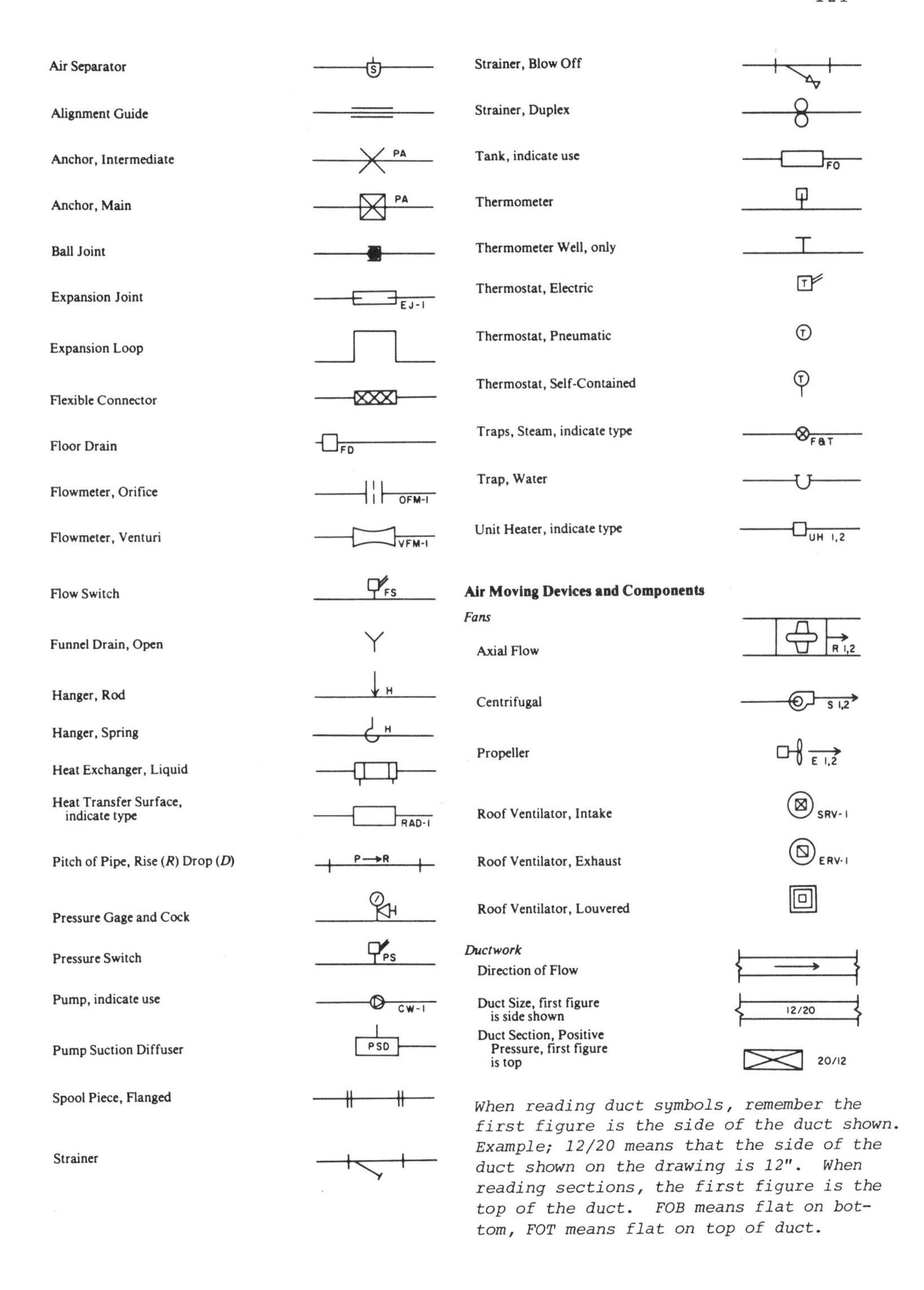

When reading duct symbols, remember the first figure is the side of the duct shown. Example; 12/20 means that the side of the duct shown on the drawing is 12". When reading sections, the first figure is the top of the duct. FOB means flat on bottom, FOT means flat on top of duct.

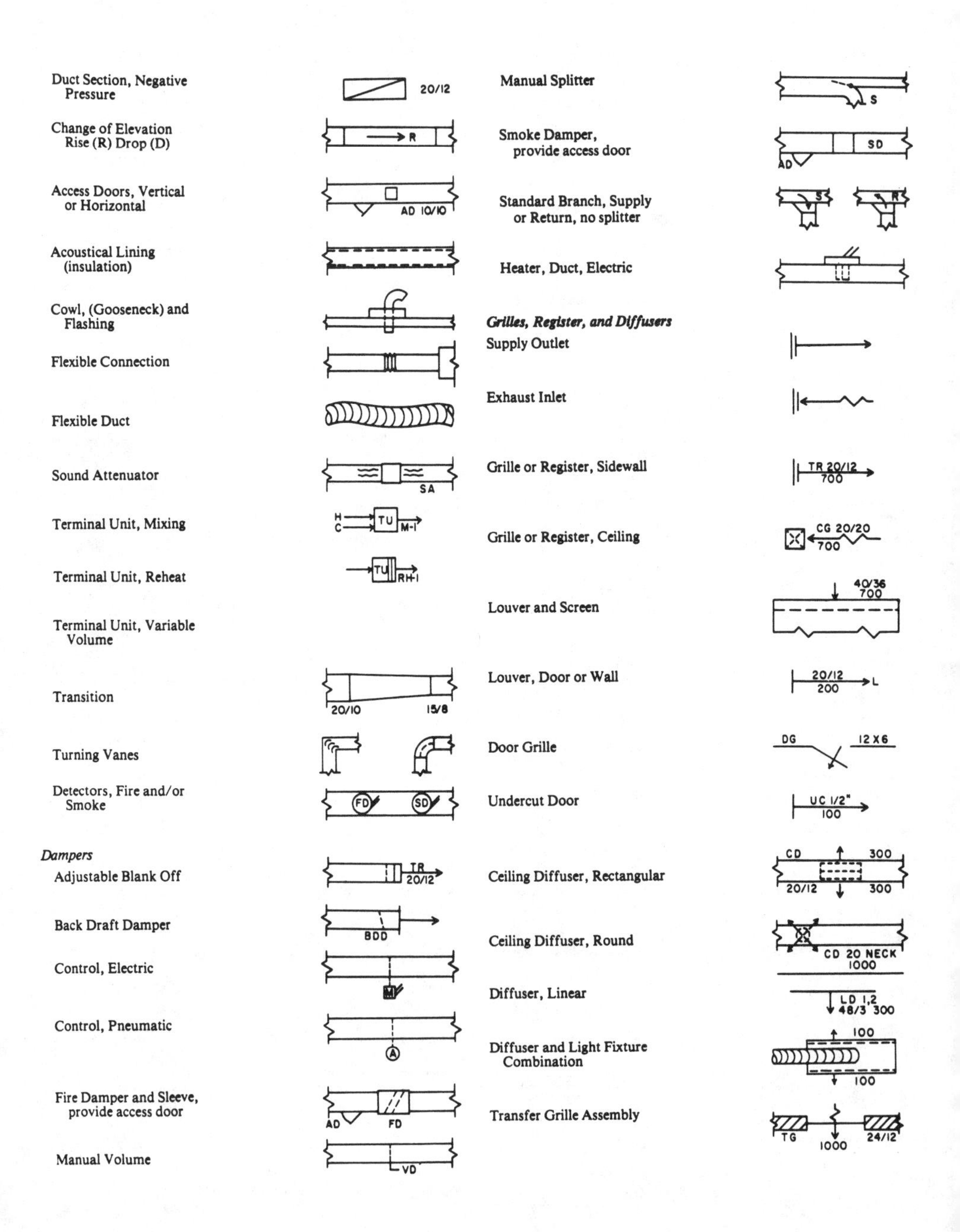
Duct Section, Negative Pressure
20/12
Change of Elevation Rise (R) Drop (D)
R
Access Doors, Vertical or Horizontal
AD 10/10
Acoustical Lining (insulation)
Cowl, (Gooseneck) and Flashing
Flexible Connection
Flexible Duct
Sound Attenuator
SA
Terminal Unit, Mixing
H
C
TU
M-1
Terminal Unit, Reheat
TU
RH-1
Terminal Unit, Variable Volume
Transition
20/10
15/8
Turning Vanes
Detectors, Fire and/or Smoke
FD
SD
Dampers
Adjustable Blank Off
TR
20/12
Back Draft Damper
BDD
Control, Electric
M
Control, Pneumatic
A
Fire Damper and Sleeve, provide access door
AD
FD
Manual Volume
VD
Manual Splitter
S
Smoke Damper, provide access door
SD
AD
Standard Branch, Supply or Return, no splitter
S
R
Heater, Duct, Electric
Grilles, Register, and Diffusers
Supply Outlet
Exhaust Inlet
Grille or Register, Sidewall
TR 20/12
700
Grille or Register, Ceiling
CG 20/20
700
Louver and Screen
40/36
700
Louver, Door or Wall
20/12
200
L
Door Grille
DG
12 X 6
Undercut Door
UC 1/2"
100
Ceiling Diffuser, Rectangular
CD
300
20/12
300
Ceiling Diffuser, Round
CD 20 NECK
1000
Diffuser, Linear
LD 1,2
48/3 300
Diffuser and Light Fixture Combination
100
100
Transfer Grille Assembly
TG
1000
24/12

Refrigeration

Compressors

Centrifugal

Reciprocating

Rotary

Rotary Screw

Condensers

Air Cooled

Evaporative

Water Cooled, specify type

Condensing Units

Air Cooled

Water Cooled

Condenser-Evaporator (Cascade System)

L.S. COND

H.S. EVAP

Cooling Towers

Cooling Tower

Spray Pond

Evaporators

Finned Coil

Forced Convection

Immersion Cooling Unit

Plate Coil

Pipe Coil

Liquid Chillers

(Chillers only)

Direct Expansion[3]

RS RL L

Flooded[3]

RS L RL

Tank, Closed

Tank, Open

Chilling Units

Absorption

L L

Centrifugal

L

Reciprocating

L

Rotary Screw

L

Controls

Refrigerant Controls

Capillary Tube

Expansion Valve, Hand

Expansion Valve, Automatic

Expansion Valve, Thermostatic

Float Valve, High Side

Float Valve, Low Side

Thermal Bulb

Solenoid Valve

S

Constant Pressure Valve, Suction

Evaporator Pressure Regulating Valve, Thermostatic, Throttling Type

Evaporator Pressure Regulating Valve, Thermostatic, Snap Action

Evaporator Pressure Regulating Valve, Throttling-Type, Evaporator Side

ES

Compressor Suction Valve, Pressure Limiting, Throttling Type, Compressor Side

CS

Thermo-Suction Valve

Snap Action Valve

Refrigerant Reversing Valve

Temperature or Temperature-Actuated Electrical or Flow Controls

Thermostat, Self-Contained

T

Thermostat, Remote Bulb

T

Pressure of Pressure-Actuated Electrical or Flow Controls

Pressure Switch

P

Pressure Switch, Dual (High-Low)

P

Pressure Switch, Differential Oil Pressure

P

Automatic Reducing Valve

Automatic Bypass Valve

Valve, Pressure Reducing

Valve, Condenser Water Regulating

Auxiliary Equipment

Refrigerant

Filter

Strainer

Filter and Drier

Scale Trap

Drier

Vibration Absorber

Heat Exchanger

Oil Separator

Sight Glass

Fusible Plug

Rupture Disc

Receiver, High Pressure, Horizontal

Receiver, High Pressure, Vertical

Receiver, Low Pressure

Intercooler

Intercooler/Desuperheater

Energy Recovery Equipment

Condenser, Double Bundle

Air to Air Energy Recovery

Rotary Heat Wheel

C H

Coil Loop

C H

Heat Pipe

C H

Fixed Plate

C H

Plate Fin, Cross Flow

C H

Power Sources

Motor, Electric, number indicates horsepower

25

COMMON DUCT SYMBOLS

SYMBOL MEANING	*SYMBOL*	*SYMBOL MEANING*	*SYMBOL*
POINT OF CHANGE IN DUCT CONSTRUCTION (BY STATIC PRESSURE CLASS)		SUPPLY GRILLE (SG)	20 × 12 SG 700 CFM
DUCT (1ST FIGURE, SIDE SHOWN 2ND FIGURE, SIDE NOT SHOWN)	20 × 12	RETURN (RG) OR EXHAUST (EG) GRILLE (NOTE AT FLR OR CLG)	20 × 12 RG 700 CFM
ACOUSTICAL LINING DUCT DIMENSIONS FOR NET FREE AREA		SUPPLY REGISTER (SR) (A GRILLE + INTEGRAL VOL. CONTROL)	20 × 12 SR 700 CFM
DIRECTION OF FLOW		EXHAUST OR RETURN AIR INLET CEILING (INDICATE TYPE)	20 × 12 GR 700 CFM
DUCT SECTION (SUPPLY)	S 30 × 12	SUPPLY OUTLET, CEILING, ROUND (TYPE AS SPECIFIED) INDICATE FLOW DIRECTION	20 700 CFM
DUCT SECTION (EXHAUST OR RETURN)	E OR R 20 × 12	SUPPLY OUTLET, CEILING, SQUARE (TYPE AS SPECIFIED) INDICATE FLOW DIRECTION	12 × 12 700 CFM
INCLINED RISE (R) OR DROP (D) ARROW IN DIRECTION OF AIR FLOW	R	TERMINAL UNIT. (GIVE TYPE AND/OR SCHEDULE)	T U
TRANSITIONS: GIVE SIZES. NOTE F.O.T. FLAT ON TOP OR F.O.B. FLAT ON BOTTOM IF APPLICABLE		COMBINATION DIFFUSER AND LIGHT FIXTURE	
		DOOR GRILLE	DG 12 × 6
STANDARD BRANCH FOR SUPPLY & RETURN (NO SPLITTER)	S R	SOUND TRAP	ST
SPLITTER DAMPER		FAN & MOTOR WITH BELT GUARD & FLEXIBLE CONNECTIONS	
VOLUME DAMPER MANUAL OPERATION	VD	VENTILATING UNIT (TYPE AS SPECIFIED)	
AUTOMATIC DAMPERS MOTOR OPERATED	SEC MOD		
ACCESS DOOR (AD) ACCESS PANEL (AP)	OR AD	UNIT HEATER (DOWNBLAST)	
FIRE DAMPER: SHOW ◄ VERTICAL POS. SHOW ◆ HORIZ. POS.	FD AD	UNIT HEATER (HORIZONTAL)	
SMOKE DAMPER	SD AD	UNIT HEATER (CENTRIFUGAL FAN) PLAN	
CEILING DAMPER OR ALTERNATE PROTECTION FOR FIRE RATED CLG	C	THERMOSTAT	T
TURNING VANES		POWER OR GRAVITY ROOF VENTILATOR-EXHAUST (ERV)	
FLEXIBLE DUCT FLEXIBLE CONNECTION		POWER OR GRAVITY ROOF VENTILATOR-INTAKE (SRV)	
GOOSENECK HOOD (COWL)		POWER OR GRAVITY ROOF VENTILATOR-LOUVERED	
BACK DRAFT DAMPER	BDD	LOUVERS & SCREEN	36 × 24L

TYPICAL DUCT CONNECTIONS

ROUND DUCT

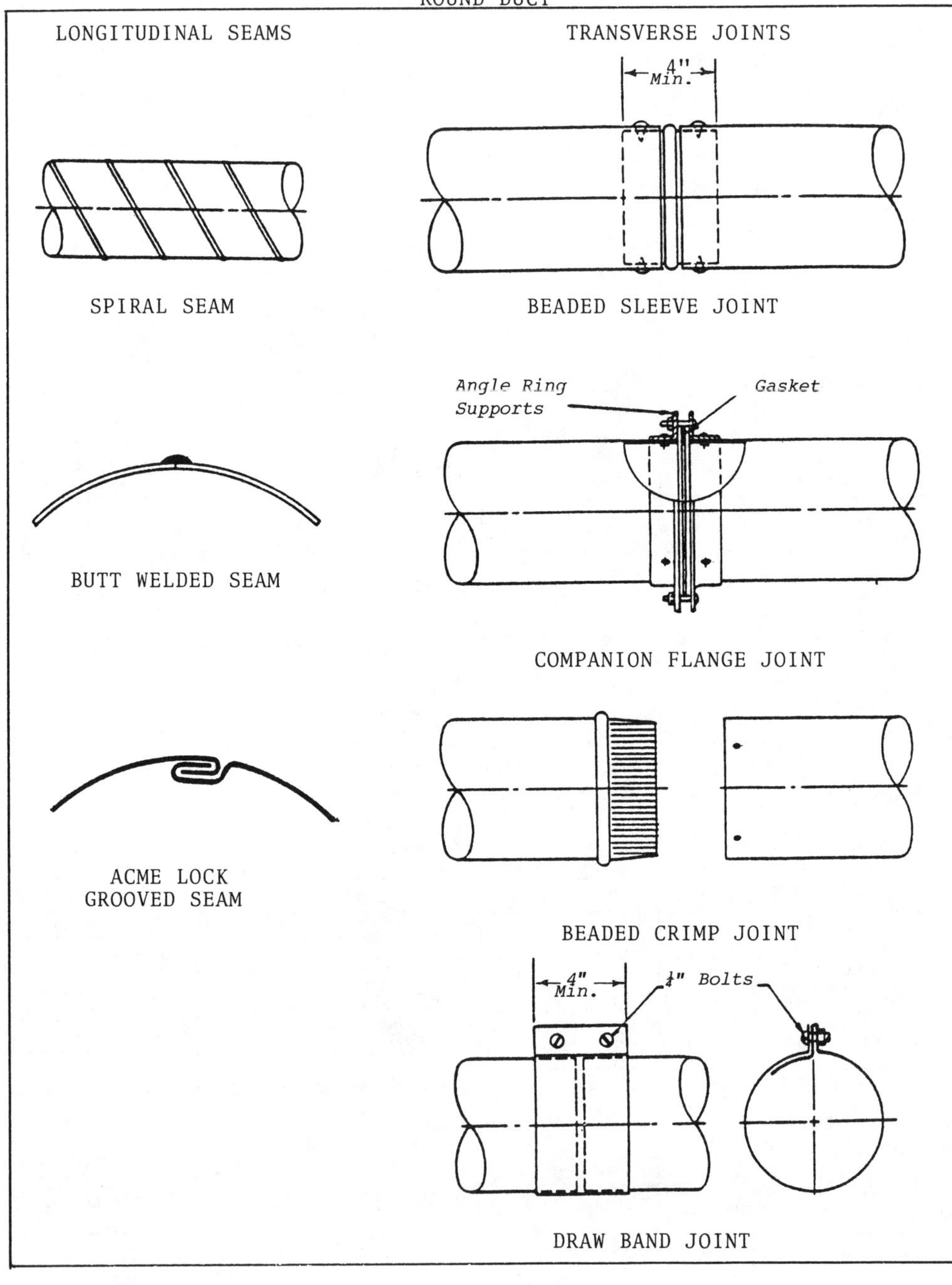

WHERE TYPICAL DUCT CONNECTIONS ARE USED

Common Joints And Seams

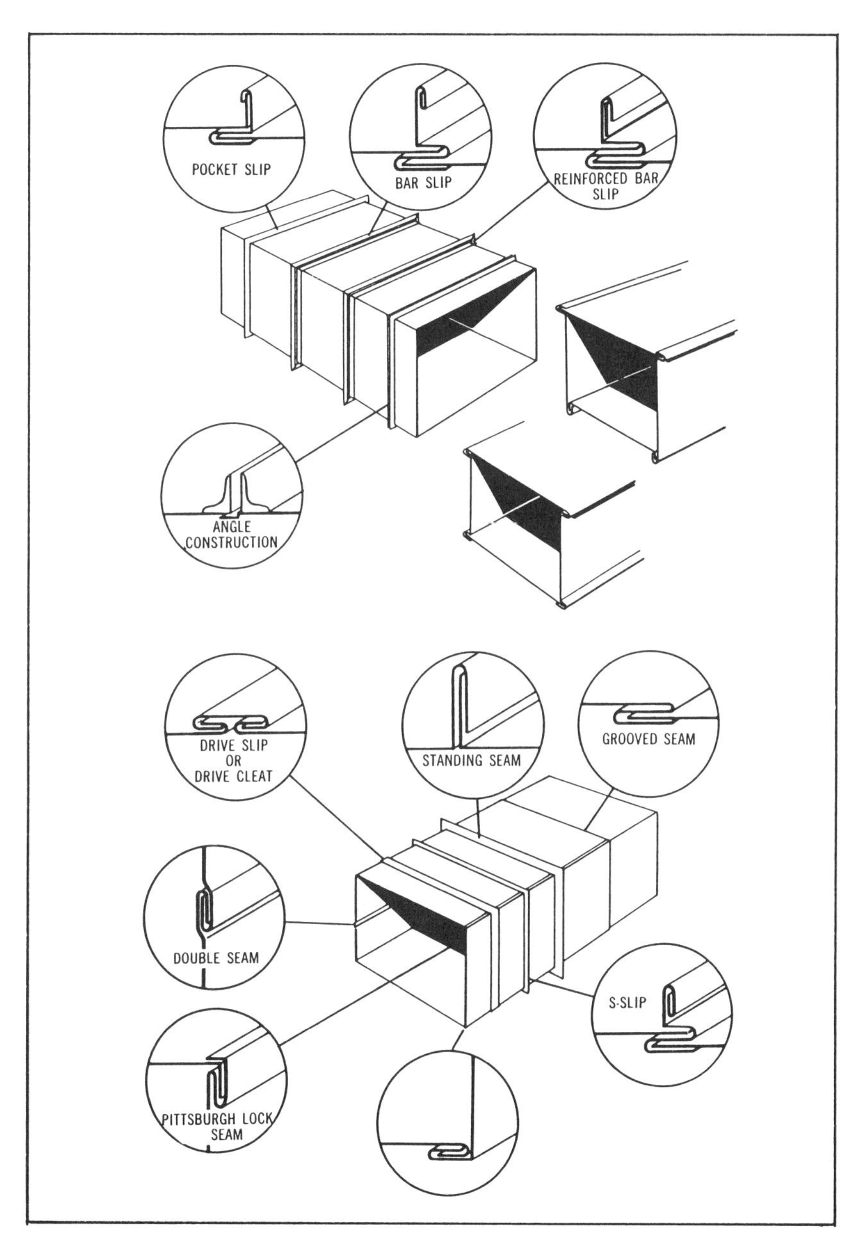

TYPICAL DUCT CONNECTIONS
CROSS JOINTS

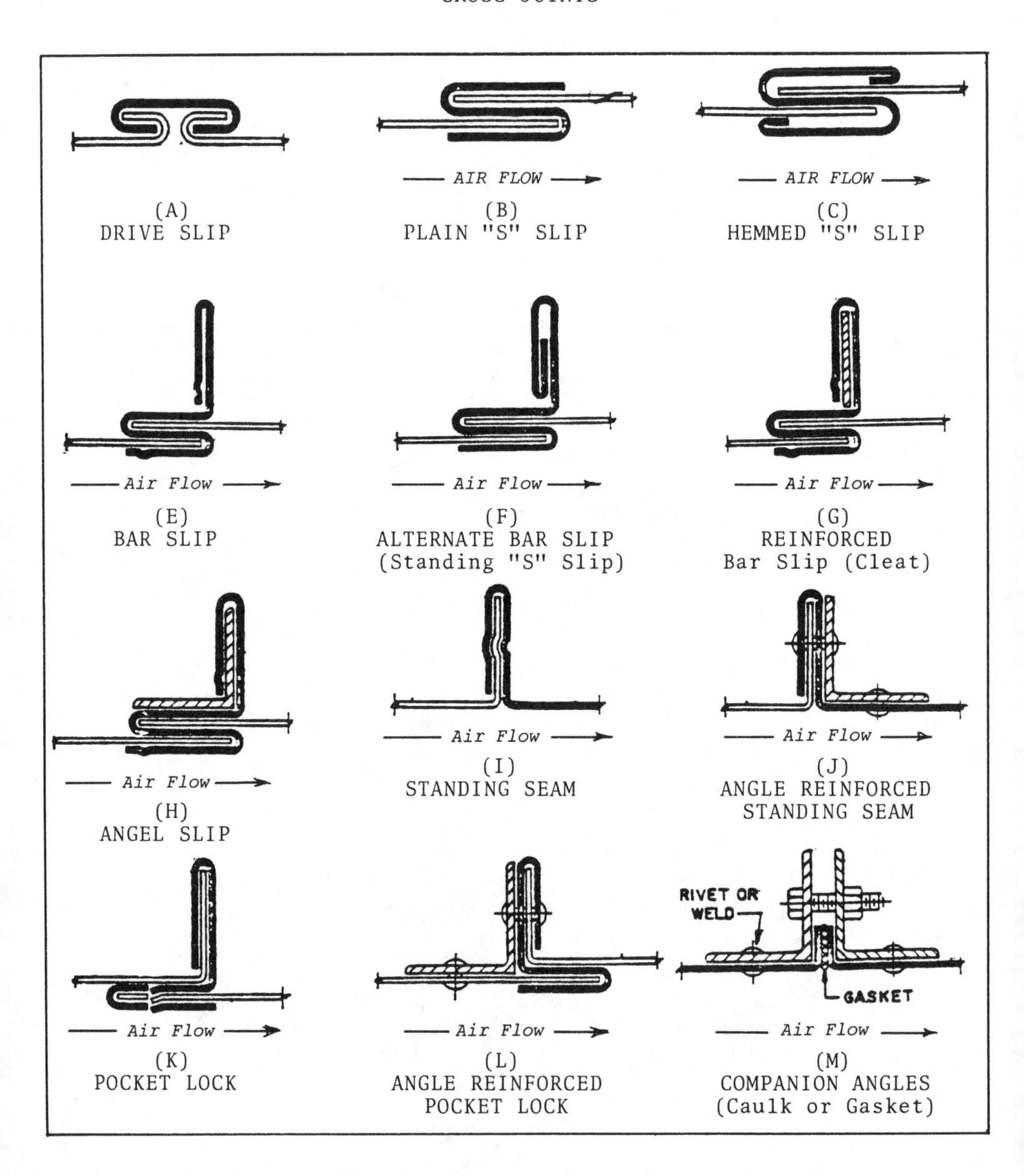

TYPICAL DUCT CONNECTIONS
LONGITUDINAL SEAMS

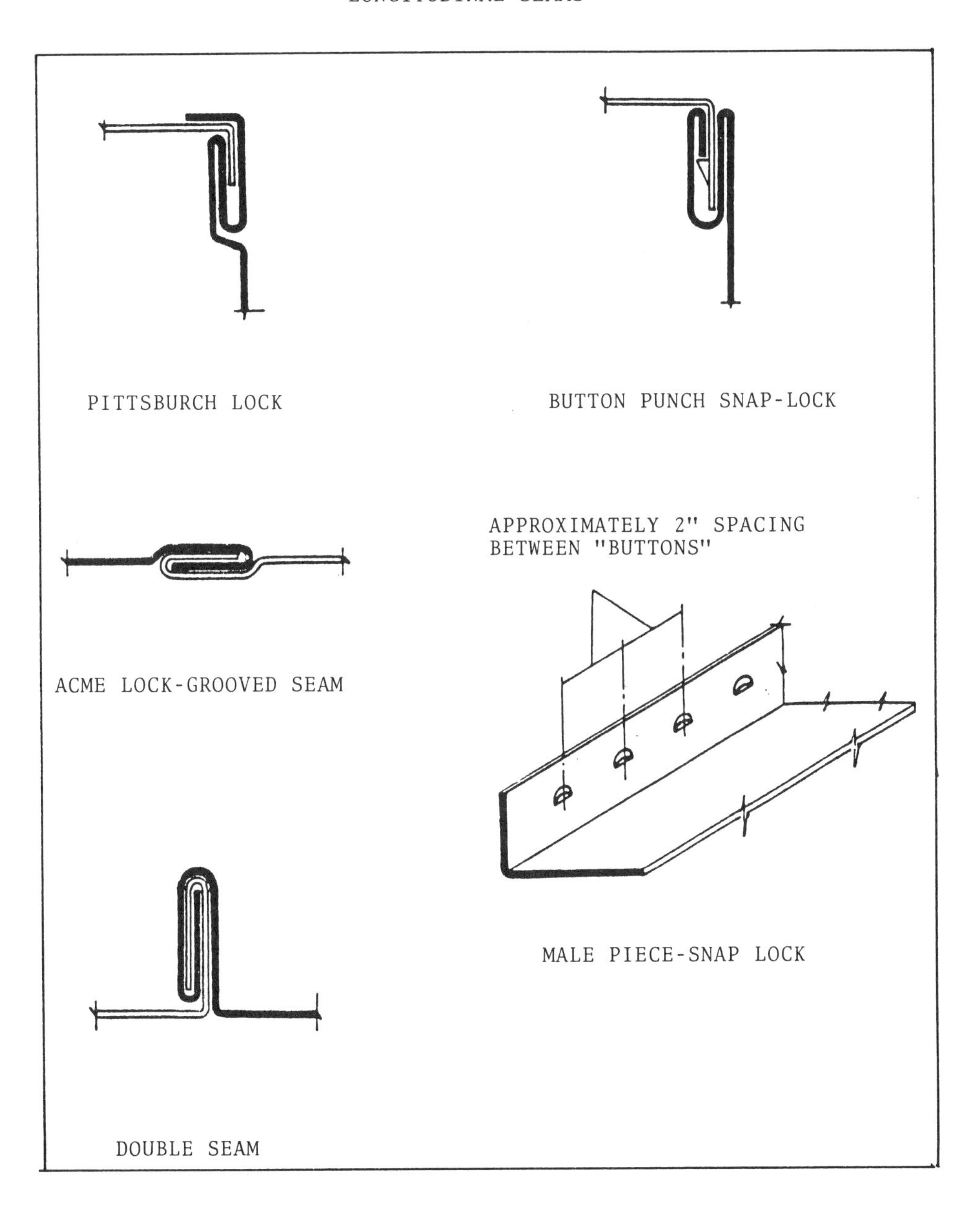

COMMON STEAM SYMBOLS

(These symbols are courtesy of Sarco. They are used in the Sarco literature. Other manufacturers may use other symbols. The exam writers often use illustrations from different manufacturer's catalogs.)

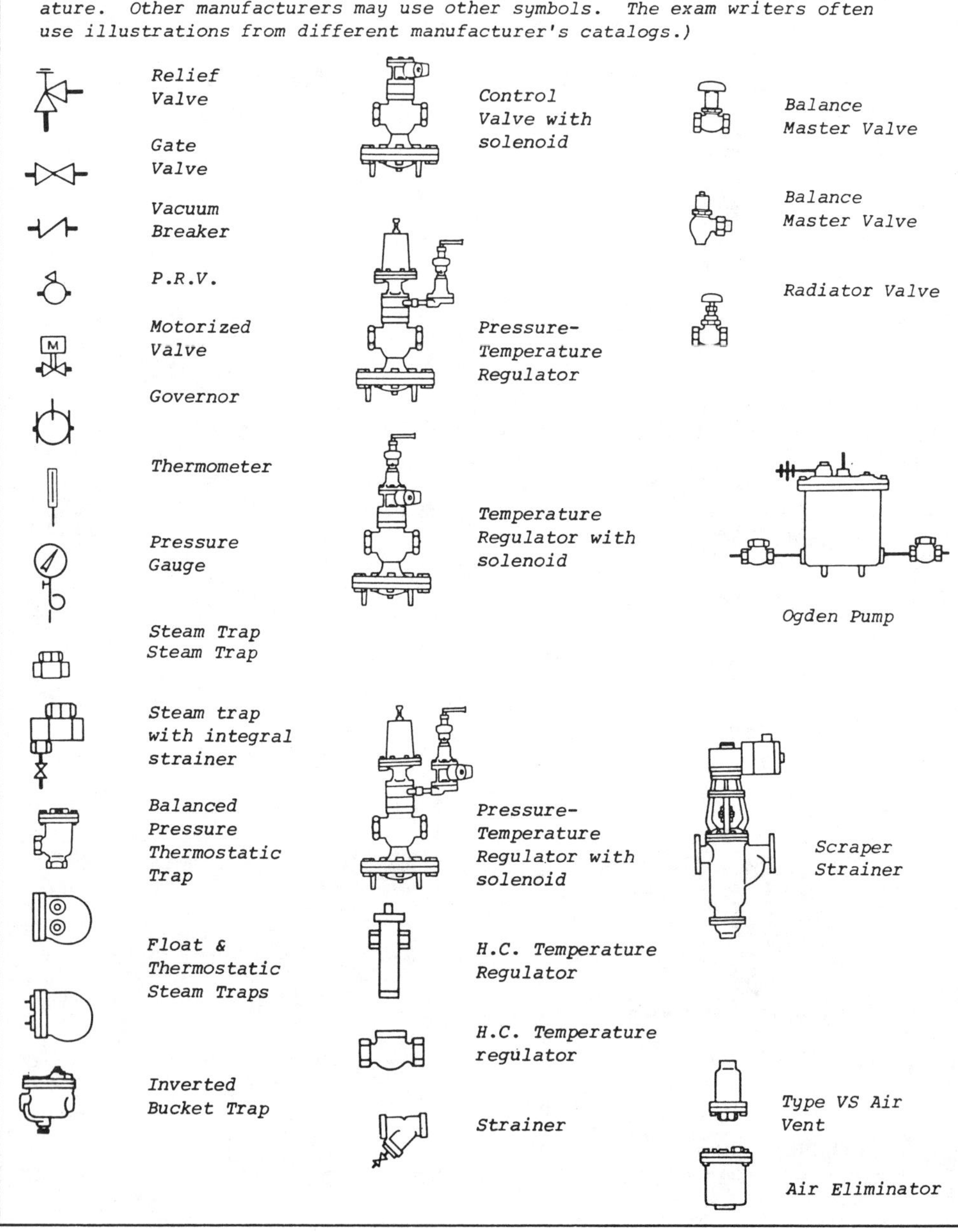

COMMON ELECTRICAL SYMBOLS

DEVICE	SYMBOL(S)	DEVICE	SYMBOL(S)
Battery		Resistance, Variable	
Capacitor (Condenser)		Switch, Single Pole, Single Throw	
Circuit Breaker (Air, Single Pole)		Switch, Double Pole Single Throw	
Coil, Nonmagnetic Core		Switch, Double Pole, Double Throw	
Coil, Magnetic Core		Pushbutton, Normally Open	
Coil for Relay		Pushbutton, Normally Closed	
Wire Crossing, No Connection		Transformer	
Wires Connected		Pilot Light	
Ground		Starter Overload (Heater)	O.L.
Contacts, Normally Open		Receptacle, Duplex	
Contacts, Normally Closed		Receptacle, Single	
Fuse		Telephone Outlet	
Ammeter	A	Incandescent Lighting Fixture	
Voltmeter	V	Fluorescent Lighting Fixture	
Wattmeter	W	Branch Circuit (3 Wire)	
Motor	M	Connection	
Resistance, Fixed (Resistor)			

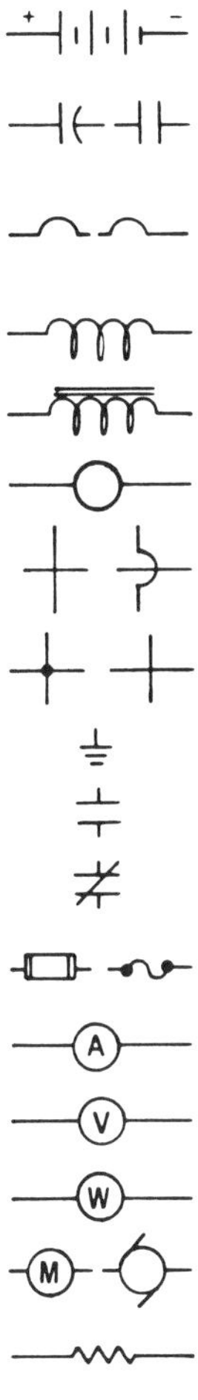

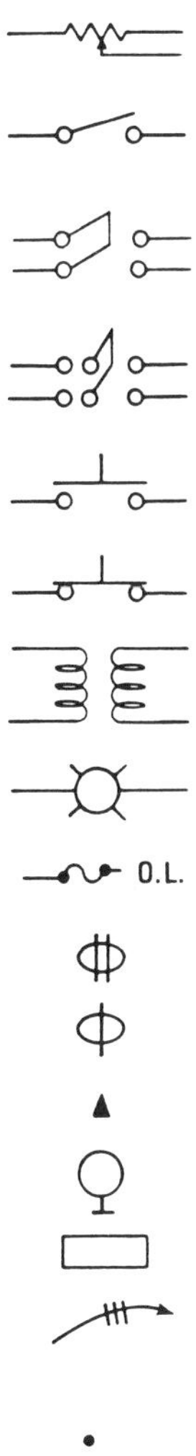

PART 3

FIELD APPLICATION PROBLEMS

SYSTEM TESTING

Refrigeration Testing

License examinations usually have questions about testing refrigeration equipment and piping systems. In all cases, the local code will override all other references. The reader is urged to become familiar with ASME B31.5-1983, § II, Table 501.2 and § VI 537.3, as well as ASHRAE 15-1989, 11.1 and Table 5. These are most frequently referenced. In areas covered by the Standard Mechanical Code, Table 4-2, 1982 or later edition prevails. For the Uniform Mechanical Code, Table 15-C.

Refrigeration Piping

Refrigeration tests shall be made at pressure not less than the minimum shown in Table 3.1 and held for 30 minutes without loss. (See ANSI/AHSRAE 15-19, 12.1 and Table 5; Standard Mechanical Code, § 600). Refrigeration piping should not be tested with water, oxygen or combustible gases. Nitrogen is generally used. For a complete discussion on testing see Modern Refrigeration and Air Conditioning, Althouse, Turnquist, Bracciano, Goodheart and Willcox, 1982, pp 361, 506-507.

Hydrostatic Testing

A piping system shall be subjected to a hydrostatic test pressure which at every point in the system is not less than 1½ times the design pressure and not greater than the maximum test pressure for any vessel, pump, valve or other component in the system under test.

Following the application of hydrostatic test pressure for at least ten minutes, examination shall be made for leakage of the piping and at all joints and connections. (See ANSI/ASME B 31.9, 937.3.4)

Boilers

All boilers shall be subjected to a hydrostatic test of 60 psi or 1½ times the maximum allowable working pressure, which ever is the greater.

Steam , Water and Brine Piping

Piping systems for brine shall be tested at 1½ times the design pressure. All pipe except cast iron and plastic shall be tested to 1½ times the operating pressure with nitrogen, CO , or compressed air (not oxygen) or hydrostatically, and held for at least 30 minutes. Cast iron shall be tested hydrostatically only. (References: ANSI/ASME B. 31.5, 537; Uniform Mechanical Code, Appendix B; Standard Mechanical Code, 609).

TABLE 3.1

MINIMUM DESIGN PRESSURES FOR REFRIGERANT TESTING

Group[a]	Refrigerant	Name	Chemical Formula	Minimum Design Gage Pressures, psi		
					High Side	
				Low Side	Water or Evaporation Cooled	Air Cooled
1	R-11	Trichlorofluoromethane	CCl_3F	15	15	21
1	R-12	Dichlorodifluoromethane	CCl_2F_2	85	127	169
1	R-13	Chlorotrifluoromethane	$CClF_3$	521	547	547
1	R-13B1	Bromotrifluoromethane	$CBrF_3$	230	321	410
1	R-14	Tetrafluoromethane	CF_4	529	529	529
1	R-21	Dichlorofluoromethane	$CHCl_2F$	15	29	46
1	R-22	Chlorodifluoromethane	$CHClF_2$	144	211	278
1	R-30	Methylene chloride	CH_2Cl_2	15	15	15
2	R-40	Methyl chloride	CH_3Cl	72	112	151
1	R-113	Trichlorotrifluoroethane	CCl_2FCClF_2	15	15	15
1	R-114	Dichlorotetrafluoroethane	$CClF_2CClF_2$	18	35	53
1	R-115	Chloropentafluoroethane	$CClF_2CF_3$	123	181	238
3	R-170	Ethane	C_2H_6	618	695	695
3	R-290	Propane	C_3H_8	129	188	244
1	R-C318	Octafluorocyclobutane	C_4F_8	34	59	85
1	R-500	Dichlorodifluoromethane, 73.8% and ethylidene fluoride, 26.2%	CCl_2F_2/CH_3CHF_2	102	153	203
1	R-502	Chlorodifluoromethane, 48.8% and chloropentafluoroethane, 51.2%	$CHClF_2/CClF_2CF_3$	162	232	302
3	R-600	N-butane	C_4H_{10}	23	42	61
3	R-601	Isobutane	$CH(CH_3)_3$	39	63	88
2	R-611	Methyl formate	$HCOOCH_3$	15	15	15
2	R-717	Ammonia	NH_3	139	215	293
1	R-744	Carbon dioxide	CO_2	955	1058	1058
2	R-764	Sulfur dioxide	SO_2	45	78	115
3	R-1150	Ethylene	C_2H_4	732	732	732

Reproduced from ASME Code for Pressure Piping, B31.5-1983, by permission American Society of Mechanical Engineers. Notes deleted.

TABLE 3.2

CLASSIFICATION OF REFRIGERANTS

Refrigerant[a,f] and Amounts[b,e]

Refrigerant Number	Chemical Name	Chemical Formula	Quantity of Refrigerant per Occupied Space: Lb per 1000 ft^3 [a,c]	PPM by vol	g/m^3 [a,c]
Group A1					
R-11	Trichlorofluoromethane	CCl_3F	1.6	4,000	25
R-12	Dichlorodifluoromethane	CCl_2F_2	12	40,000	200
R-13	Chlorotrifluoromethane	$CClF_3$	18	67,000	290
R-13B1	Bromotrifluoromethane	$CBrF_3$	22	57,000	350
R-14	Tetrafluoromethane (Carbon tetrafluoride)	CF_4	15	67,000	240
R-22	Chlorodifluoromethane	$CHClF_2$	9.4	42,000	150
R-113	Trichlorotrifluoroethane	CCl_2FCClF_2	1.9	4,000	31
R-114	Dichlorotetrafluoroethane	$CClF_2CClF_2$	9.4	21,000	150
R-115	Chloropentafluoroethane	$CClF_2CF_3$	27	67,000	430
R-134a	1,1,1,2-Tetrafluoroethane	CH_2FCF_3	16	60,000	250
R-C318	Octafluorocyclobutane	C_4F_8	35	67,000	550
R-400	R-12 and R-114	$CCl_2F_2/C_2Cl_2F_4$	d	d	d
R-500	R-12/152a (73.8/26.2)	CCl_2F_2/CH_3CHF_2	12	47,000	200
R-502	R-22/115 (48.8/51.2)	$CHClF_2/CClF_2CF_3$	19	65,000	300
R-503	R-23/13 (40.1/59.9)	$CHF_3/CClF_3$	15	67,000	240
R-744	Carbon Dioxide	CO_2	5.7	50,000	91
Group A2					
R-142b	1-Chloro-1,1,-Difluoroethane	CH_3CClF_2	3.7	14,000	60
R-152a	1,1-Difluoroethane	CH_3CHF_2	1.2	7,000	20
Group A3					
R-170	Ethane	C_2H_6	0.50	6,400	8.0
R-290	Propane	C_3H_8	0.50	4,400	8.0
R-600	Butane	C_4H_{10}	0.51	3,400	8.2
R-600a	2-Methyl propane (Isobutane)	$CH(CH_3)_3$	0.51	3,400	8.2
R-1150	Ethene (Ethylene)	C_2H_4	0.38	5,200	6.0
R-1270	Propene (Propylene)	C_3H_6	0.37	3,400	5.9
Group B1					
R-123	2,2-Dichloro-1,1,1-Trifluoroethane	$CHCl_2CF_3$	0.40	1,000	6.3
R-764	Sulfur Dioxide	SO_2	0.016	100	0.26
Group B2					
R-40	Chloromethane (Methyl Chloride)	CH_3Cl	1.3	10,000	21
R-611	Methyl Formate	$HCOOCH_3$	0.78	5,000	12
R-717	Ammonia	NH_3	0.022	500	0.35

[a]The refrigerant safety groups in Table 1 are not part of ASHRAE Standard 15. The classifications shown are from ASHRAE Standard 34, which governs in the event of a difference.

[b]To be used only in conjunction with Section 7.

[c]To correct for height, H(ft), above sea level, multiply these values by $(1 - 2.42 \times 10^{-6}H)$. To correct for height, h(km), above sea level, multiply these values by $(1 - 7.94 \times 10^{-2}h)$.

[d]The quantity of each component shall comply with the limits set in Table 1 for the pure compound, and the total volume % of all components shall be calculated per Appendix A (not to exceed 67,000 ppm by volume for any refrigerant blend).

[e]The basis of the table quantities is a single event where a complete discharge of any refrigerant system into the occupied space occurs. The quantity of refrigerant is the most restrictive of a minimum oxygen concentration of 19.5% or as follows:

- Group A1— 80% of the cardiac sensitization level for R-11, R-12, R-13B1, R-22, R-113, R-114, R-134a, R-500, and R-502. 100% of the IDLH (21) for R-744. Others are limited by levels where oxygen deprivation begins to occur.
- Groups A2, A3— Approximately 20% of LFL.
- Group B1— 100% of IDLH for R-764, and 100% of the measure consistent with the IDLH for R-123.
- Groups B2, B3— 100% of IDLH or 20% of LFL, whichever is lower.

[f]It shall be the responsibility of the owner to establish the refrigerant group for refrigerants used that are not classified in ASHRAE Standard 34.

TABLE 3.3

REFRIGERANT PIPE SIZE SELECTOR--SPEED SHEET

For rapid estimating of refrigerant line sizes for average air-conditioning applications based on 40°F suction and 105°F condensing temperatures at average pressure loss 2°F.

R-12

EQUIVALENT LENGTH OF PIPE IN FEET

TONS	20 FT.			30 FT.			40 FT.			50 FT.			60 FT.			70 FT.			80 FT.			90 FT.			100 FT.		
	LIQ.	SUCT.	DISCH.	LIQ.	SUCT.	DISCH.	LIQ.	SUCT.	DISCH.	LIQ.	SUCT.	DISCH.	LIQ.	SUCT.	DISCH.	LIQ.	SUCT.	DISCH.	LIQ.	SUCT.	DISCH.	LIQ.	SUCT.	DISCH.	LIQ.	SUCT.	DISCH.
2	1/2	3/4	5/8	1/2	7/8	5/8	1/2	7/8	3/4	1/2	7/8	3/4	1/2	1 1/8	3/4	1/2	1 1/8	3/4	1/2	1 1/8	3/4	1/2	1 1/8	3/4	1/2	1 1/8	3/4
3	1/2	7/8	3/4	1/2	1 1/8	3/4	1/2	1 1/8	3/4	1/2	1 1/8	3/4	5/8	1 1/8	3/4	5/8	1 1/8	7/8	5/8	1 1/8	7/8	5/8	1 1/8	7/8	5/8	1 1/8	7/8
5	1/2	1 1/8	7/8	5/8	1 1/8	7/8	5/8	1 1/8	7/8	5/8	1 3/8	1 1/8	5/8	1 3/8	1 1/8	3/4	1 3/8	1 1/8	5/8	1 3/8	1 1/8	3/4	1 3/8	1 1/8	3/4	1 3/8	1 1/8
8	5/8	1 3/8	1 1/8	5/8	1 3/8	1 1/8	3/4	1 3/8	1 1/8	3/4	1 5/8	1 1/8	3/4	1 5/8	1 1/8	3/4	1 5/8	1 3/8	7/8	1 5/8	1 3/8	7/8	1 5/8	1 3/8	7/8	2 1/8	1 3/8
10	5/8	1 3/8	1 1/8	3/4	1 5/8	1 1/8	3/4	1 5/8	1 1/8	3/4	1 5/8	1 3/8	7/8	1 5/8	1 3/8	7/8	1 5/8	1 3/8	7/8	2 1/8	1 3/8	7/8	2 1/8	1 3/8	7/8	2 1/8	1 3/8
15	3/4	1 5/8	1 3/8	7/8	1 5/8	1 3/8	7/8	2 1/8	1 3/8	7/8	2 1/8	1 3/8	1 1/8	2 1/8	1 5/8	1 1/8	2 1/8	1 5/8	1 1/8	2 1/8	1 5/8	1 1/8	2 1/8	1 5/8	1 1/8	2 1/8	1 5/8
20	7/8	1 5/8	1 3/8	7/8	2 1/8	1 3/8	1 1/8	2 1/8	1 5/8	1 1/8	2 1/8	1 5/8	1 1/8	2 1/8	1 5/8	1 1/8	2 1/8	1 5/8	1 1/8	2 5/8	2 1/8	1 1/8	2 5/8	2 1/8	1 1/8	2 5/8	2 1/8
25	7/8	2 1/8	1 5/8	1 1/8	2 1/8	1 5/8	1 1/8	2 1/8	1 5/8	1 1/8	2 5/8	1 5/8	1 1/8	2 5/8	2 1/8	1 1/8	2 5/8	2 1/8	1 1/8	2 5/8	2 1/8	1 3/8	2 5/8	2 1/8	1 3/8	2 5/8	2 1/8
30	1 1/8	2 1/8	1 5/8	1 1/8	2 1/8	1 5/8	1 1/8	2 5/8	2 1/8	1 1/8	2 5/8	2 1/8	1 1/8	2 5/8	2 1/8	1 3/8	2 5/8	2 1/8	1 3/8	2 5/8	2 1/8	1 3/8	2 5/8	2 1/8	1 3/8	2 5/8	2 1/8
40	1 1/8	2 1/8	1 5/8	1 1/8	2 5/8	2 1/8	1 3/8	2 5/8	2 1/8	1 3/8	2 5/8	2 1/8	1 3/8	2 5/8	2 1/8	1 3/8	3 1/8	2 1/8	1 3/8	3 1/8	2 1/8	1 3/8	3 1/8	2 5/8	1 5/8	3 1/8	2 5/8
50	1 1/8	2 5/8	2 1/8	1 3/8	2 5/8	2 1/8	1 3/8	2 5/8	2 1/8	1 3/8	3 1/8	2 1/8	1 3/8	3 1/8	2 5/8	1 5/8	3 1/8	2 5/8	1 5/8	3 1/8	2 5/8	1 5/8	3 5/8	2 5/8	1 5/8	3 5/8	2 5/8
60	1 3/8	2 5/8	2 1/8	1 3/8	3 1/8	2 1/8	1 3/8	3 1/8	2 5/8	1 5/8	3 1/8	2 5/8	1 5/8	3 1/8	2 5/8	1 5/8	3 5/8	2 5/8	1 5/8	3 5/8	2 5/8	1 5/8	3 5/8	2 5/8	1 5/8	3 5/8	2 5/8
70	1 3/8	2 5/8	2 1/8	1 3/8	3 1/8	2 5/8	1 5/8	3 1/8	2 5/8	1 5/8	3 5/8	2 5/8	1 5/8	3 5/8	2 5/8	1 5/8	3 5/8	2 5/8	2 1/8	3 5/8	2 5/8	2 1/8	3 5/8	3 1/8	2 1/8	4 1/8	3 1/8
80	1 3/8	3 1/8	2 1/8	1 5/8	3 1/8	2 5/8	1 5/8	3 5/8	2 5/8	1 5/8	3 5/8	2 5/8	1 5/8	3 5/8	2 5/8	2 1/8	3 5/8	3 1/8	2 1/8	4 1/8	3 1/8	2 1/8	4 1/8	3 1/8	2 1/8	4 1/8	3 1/8
90	1 3/8	3 1/8	2 5/8	1 5/8	3 1/8	2 5/8	1 5/8	3 5/8	2 5/8	1 5/8	3 5/8	2 5/8	2 1/8	3 5/8	3 1/8	2 1/8	4 1/8	3 1/8	2 1/8	4 1/8	3 1/8	2 1/8	4 1/8	3 1/8	2 1/8	4 1/8	3 1/8
100	1 5/8	3 1/8	2 5/8	1 5/8	3 5/8	2 5/8	1 5/8	3 5/8	2 5/8	2 1/8	3 5/8	3 1/8	2 1/8	4 1/8	3 1/8	2 1/8	4 1/8	3 1/8	2 1/8	4 1/8	3 1/8	2 1/8	4 1/8	3 1/8	2 1/8	5 1/8	3 1/8

CONDENSATE LINE SIZING, CONDENSER TO RECEIVER

CONDENSATE LINE SIZE	1/2	5/8	3/4	7/8	1 1/8	1 3/8	1 5/8	2 1/8	2 5/8	3 1/8	3 5/8	4 1/8
REFRIGERATION MAXIMUM TONS	1.5	3.0	5.0	7.5	14.0	23.0	35.0	70.0	110.0	170.0	245.0	330.0
MINIMUM HEIGHT	10"	10"	10"	10"	14"	14"	14"	16"	16"	16"	16"	16"

R-22

EQUIVALENT LENGTH OF PIPE IN FEET

TONS	20 FT.			30 FT.			40 FT.			50 FT.			60 FT.			70 FT.			80 FT.			90 FT.			100 FT.		
	LIQ.	SUCT.	DISCH.	LIQ.	SUCT.	DISCH.	LIQ.	SUCT.	DISCH.	LIQ.	SUCT.	DISCH.	LIQ.	SUCT.	DISCH.	LIQ.	SUCT.	DISCH.	LIQ.	SUCT.	DISCH.	LIQ.	SUCT.	DISCH.	LIQ.	SUCT.	DISCH.
2	3/8	5/8	1/2	3/8	5/8	1/2	3/8	3/4	1/2	3/8	3/4	5/8	3/8	3/4	5/8	1/2	3/4	5/8	1/2	3/4	5/8	1/2	3/4	5/8	1/2	7/8	5/8
3	3/8	3/4	5/8	3/8	3/4	5/8	1/2	3/4	5/8	1/2	7/8	5/8	1/2	7/8	5/8	1/2	7/8	3/4	1/2	7/8	3/4	1/2	7/8	3/4	1/2	7/8	3/4
5	1/2	7/8	5/8	1/2	7/8	5/8	1/2	1 1/8	3/4	1/2	1 1/8	3/4	5/8	1 1/8	3/4	5/8	1 1/8	7/8	5/8	1 1/8	7/8	5/8	1 1/8	7/8	5/8	1 1/8	7/8
8	1/2	1 1/8	3/4	5/8	1 1/8	3/4	5/8	1 1/8	7/8	5/8	1 1/8	7/8	5/8	1 1/8	1 1/8	5/8	1 3/8	1 1/8	3/4	1 3/8	1 1/8	3/4	1 3/8	1 1/8	3/4	1 3/8	1 1/8
10	5/8	1 1/8	7/8	5/8	1 1/8	3/4	5/8	1 1/8	7/8	5/8	1 3/8	1 1/8	3/4	1 3/8	1 1/8	3/4	1 3/8	1 1/8	3/4	1 3/8	1 1/8	3/4	1 3/8	1 1/8	3/4	1 3/8	1 1/8
15	5/8	1 3/8	1 1/8	3/4	1 3/8	1 1/8	3/4	1 3/8	1 1/8	3/4	1 3/8	1 1/8	3/4	1 5/8	1 1/8	7/8	1 5/8	1 1/8	7/8	1 5/8	1 3/8	7/8	1 5/8	1 3/8	7/8	1 5/8	1 3/8
20	3/4	1 3/8	1 1/8	3/4	1 3/8	1 1/8	7/8	1 5/8	1 1/8	7/8	1 5/8	1 3/8	7/8	1 5/8	1 3/8	7/8	1 5/8	1 3/8	7/8	2 1/8	1 3/8	1 1/8	2 1/8	1 3/8	1 1/8	2 1/8	1 3/8
25	3/4	1 5/8	1 1/8	7/8	1 5/8	1 1/8	7/8	1 5/8	1 3/8	7/8	2 1/8	1 3/8	1 1/8	2 1/8	1 3/8	1 1/8	2 1/8	1 3/8	1 1/8	2 1/8	1 3/8	1 1/8	2 1/8	1 5/8	1 1/8	2 1/8	1 5/8
30	7/8	1 5/8	1 3/8	7/8	1 5/8	1 3/8	7/8	2 1/8	1 3/8	1 1/8	2 1/8	1 3/8	1 1/8	2 1/8	1 5/8	1 1/8	2 1/8	1 5/8	1 1/8	2 1/8	1 5/8	1 1/8	2 1/8	1 5/8	1 1/8	2 1/8	1 5/8
40	7/8	2 1/8	1 3/8	1 1/8	2 1/8	1 3/8	1 1/8	2 1/8	1 5/8	1 1/8	2 1/8	1 5/8	1 1/8	2 1/8	1 5/8	1 1/8	2 1/8	1 5/8	1 1/8	2 5/8	1 5/8	1 3/8	2 5/8	2 1/8	1 3/8	2 5/8	2 1/8
50	1 1/8	2 1/8	1 3/8	1 1/8	2 1/8	1 5/8	1 1/8	2 1/8	1 5/8	1 1/8	2 5/8	1 5/8	1 3/8	2 5/8	2 1/8	1 3/8	2 5/8	2 1/8	1 3/8	2 5/8	2 1/8	1 3/8	2 5/8	2 1/8	1 3/8	2 5/8	2 1/8
60	1 1/8	2 1/8	1 5/8	1 1/8	2 1/8	1 5/8	1 1/8	2 5/8	2 1/8	1 3/8	2 5/8	2 1/8	1 3/8	2 5/8	2 1/8	1 3/8	2 5/8	2 1/8	1 3/8	2 5/8	2 1/8	1 3/8	2 5/8	2 1/8	1 3/8	3 1/8	2 1/8
70	1 1/8	2 1/8	1 5/8	1 1/8	2 5/8	2 1/8	1 3/8	2 5/8	2 1/8	1 3/8	2 5/8	2 1/8	1 3/8	2 5/8	2 1/8	1 3/8	2 5/8	2 1/8	1 3/8	3 1/8	2 1/8	1 5/8	3 1/8	2 1/8	1 5/8	3 1/8	2 1/8
80	1 1/8	2 5/8	1 5/8	1 3/8	2 5/8	2 1/8	1 3/8	2 5/8	2 1/8	1 3/8	2 5/8	2 1/8	1 3/8	3 1/8	2 1/8	1 5/8	3 1/8	2 1/8	1 5/8	3 1/8	2 1/8	1 5/8	3 1/8	2 5/8	1 5/8	3 1/8	2 5/8
90	1 1/8	2 5/8	2 1/8	1 3/8	2 5/8	2 1/8	1 3/8	2 5/8	2 1/8	1 3/8	3 1/8	2 1/8	1 5/8	3 1/8	2 1/8	1 5/8	3 1/8	2 5/8	1 5/8	3 1/8	2 5/8	1 5/8	3 1/8	2 5/8	1 5/8	3 1/8	2 5/8
100	1 3/8	2 5/8	2 1/8	1 3/8	2 5/8	2 1/8	1 3/8	3 1/8	2 1/8	1 5/8	3 1/8	2 1/8	1 5/8	3 1/8	2 5/8	1 5/8	3 1/8	2 5/8	1 5/8	3 1/8	2 5/8	1 5/8	3 1/8	2 5/8	1 5/8	3 5/8	2 5/8

CONDENSATE LINE SIZING, CONDENSER TO RECEIVER

CONDENSATE LINE SIZE	1/2	5/8	3/4	7/8	1 1/8	1 3/8	1 5/8	2 1/8	2 5/8	3 1/8	3 5/8	4 1/8
REFRIGERATION MAXIMUM TONS	1.8	3.7	6.0	9.0	17.0	28.0	42.0	84.0	131.0	205.0	295.0	395.0

HOW TO USE THE CHARGING AND TESTING MANIFOLD EFFECTIVELY

The following discussion is copyrighted by Henry Valve Co., used by permission.

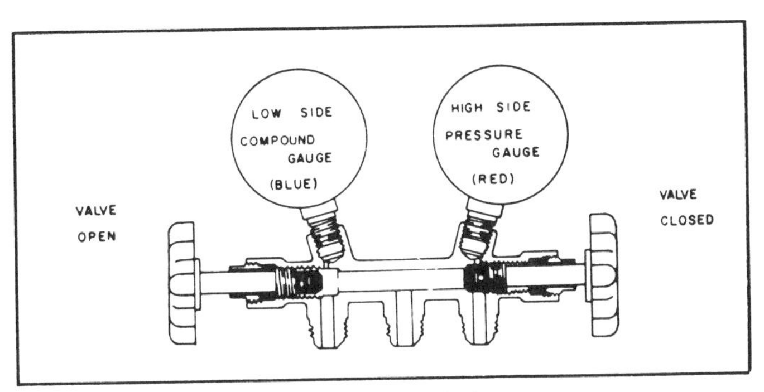

FIGURE 3.1

The charging and testing manifold is one of the most useful instruments the air conditioning and refrigeration service engineer has. With it one can perform a variety of tests to determine how a refrigerant system is performing. The scope of the tests capable of being performed range from checking the pressures in an operating system to aiding in the evacuation and charging of a system. Some of the test procedures possible with the charging and testing manifold will be shown later. Many other tests which are possible will be apparent once one becomes thoroughly familiar with the equipment.

Manifold design

First, however, a basic understanding of the manifold's design and its features will enable the operating or service engineer to use this instrument to its fullest capabilities. Fig. 1 is a cutaway drawing of the manifold, showing the gauges installed.

The manifold body is manufactured of forged brass which provides extra strength and durability. It is designed with full ports throughout which reduce evacuation and charging time to a minimum.

On both high and low sides, the gauge is open to the port at all times. The center port opening is controlled by the position of the manifold valves.

The use of color coded charging lines is suggested to avoid confusion and to facilitate line tracing. The recommended usage on these hoses is:

RED—High side pressure line
WHITE—Charging or purging line
BLUE—Low side pressure or vacuum line

When hoses aren't in use, cap ends or use hose holder to keep lines clean and dirt free.

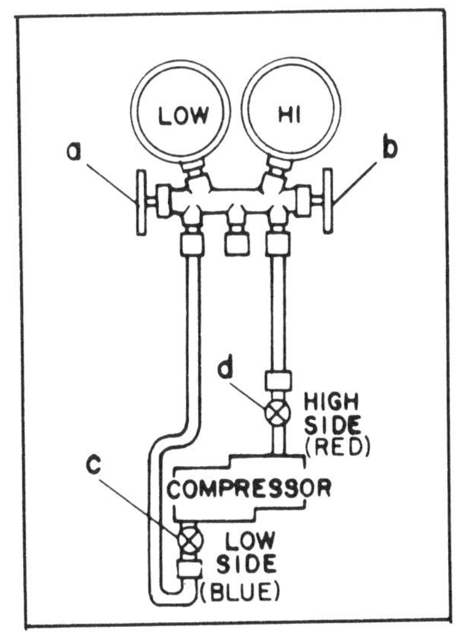

FIGURE 3.2

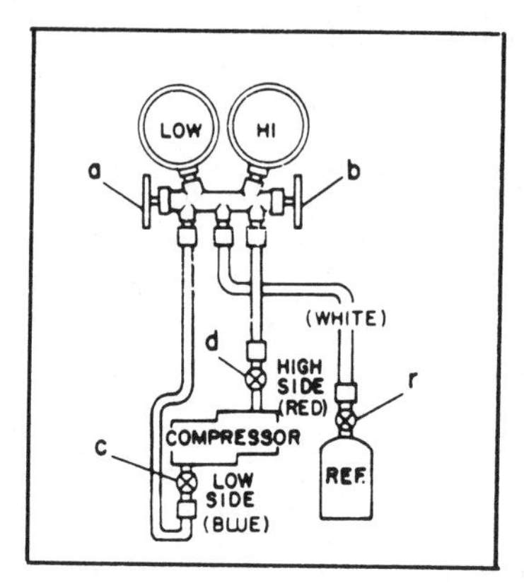

FIGURE 3.3

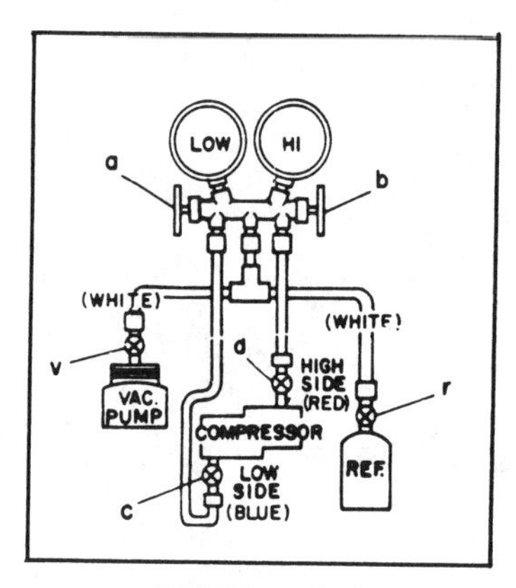

FIGURE 3.4

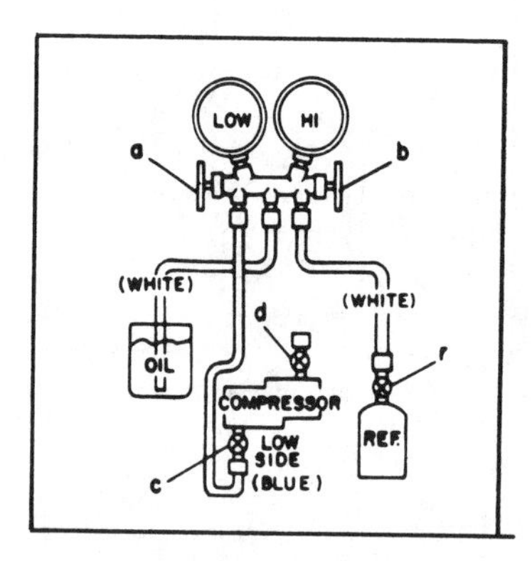

FIGURE 3.5

Operating procedures

The following are typical procedures that can be followed using the testing manifold.

It is recommended that the service engineer consult the compressor manufacturer's literature for any specific information regarding testing procedures.

1. To Observe Operating Pressures (Fig. 2)
   a. Backseat service valves *c* and *d*.
   b. Close manifold valves *a* and *b*.
   c. Connect color coded hoses as shown.
   d. Move service valves *c* and *d* off the backseat.
   e. Purge hoses at manifold.
   f. Start unit and allow to run at least five minutes.
   g. Read operating pressures or corresponding temperatures on gauges.

2. To Purge Refrigerant From System (Fig. 2)
   a. Be sure system is turned off.
   b. Close manifold valves *a* and *b*.
   c. Connect color coded hoses as shown.
   d. Service valves *c* and *d* should be open but not backseated.
   e. Slowly open manifold valves *a* and *b*. Purge through center port.
   f. When gauges read 0 lbs. pressure, purging operation is complete.

3. To Add Vapor Refrigerant Through Suction Service Valve (Fig. 3)
   a. Close manifold valves *a* and *b* and turn system off.
   b. Backseat service valves *c* and *d*.
   c. Connect color coded hoses as shown.
   d. Open refrigerant cylinder valve *r* and manifold valves *a* and *b*. Purge hoses at service valves *c* and *d*.
   e. Close manifold valves *a* and *b*.
   f. Move high side service valve *d* off backseat.
   g. Open low side service valve *c* about half way.
   h. Start unit.
   i. Meter in correct amount of refrigerant by opening and closing low side manifold valve *a*.

4. To Evacuate and Vapor Charge a System . . . if system is running . . .
   a. Observe operating pressure (Procedure 1).
   b. Turn system off.
   c. Purge system (Procedure 2).

   To Evacuate the System (Fig. 4)

   a. Close manifold valves *a* and *b*.
   b. Connect color coded hoses as shown.
   c. Be sure the service valves *c* and *d* are open but not backseated.
   d. The refrigerant cylinder valve *r* should be closed.
   e. Open manifold valves *a* and *b*.
   f. Open vacuum pump valve *v* and start vacuum pump.

   When Desired Vacuum is Reached:

   a. Close manifold valves *a* and *b*.
   b. Close vacuum pump valve *v*.
   c. Stop vacuum pump.

   To Break the Vacuum and Charge:

   a. Open refrigerant cylinder valve *r*.
   b. Low side service valve *c* should be about half way open.
   c. Start refrigeration unit.

d. Meter in correct amount of refrigerant by opening and closing low side manifold valve *a*.

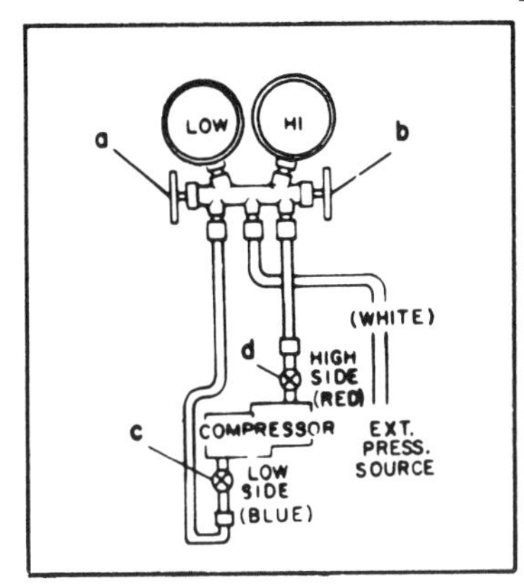

FIGURE 3.6

5. To Charge Oil Through Suction Service Valve (Fig. 5)

a. Close manifold valves *a* and *b*. Make sure unit is turned off.
b. Connect color coded hoses as shown.
c. Oil container should be filled with enough oil to fill the compressor plus enough oil to insure that the hose opening remains immersed.
d. Open refrigerant cylinder valve *r*.
e. Open manifold valves *a* and *b*. Purge air at low side service valve *c* and at oil reservoir.
f. Close manifold valves *a* and *b*.
g. Close refrigerant cylinder valve *r*.
h. Front seat low side service valve *c*.
i. Start unit and allow to run to build up low side vacuum.

Caution: Hermetic type compressors should not be operated over 18" hg vacuum to avoid possible damage to motor windings.

j. Turn unit off.
k. Meter in correct amount of dry refrigeration grade compressor oil, by opening and closing low side manifold valve *a*.

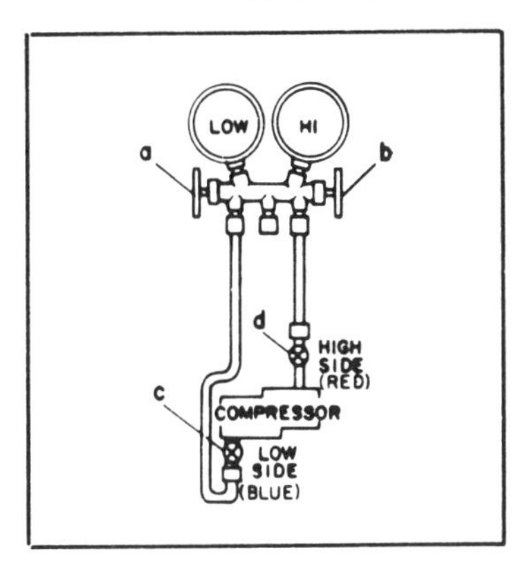

FIGURE 3.7

6. To Test Condition of Compressor (reed) Valves.
High Side (Fig. 6)

a. Be sure system is not operating.
b. Close manifold valves *a* and *b*.
c. Purge all refrigerant from system and/or compressor.
d. Connect color coded hoses as shown.
e. Front seat service valves *c* and *d*.
f. Open high side manifold valve *b* and apply pressure to compressor discharge valve.
g. Read and record pressures.
h. Close manifold valve *b*.
i. Wait several minutes.
j. Reread high side pressure. If valves are good, readings will not change appreciably.

7. Alternate Method—If System is Operating (Fig. 7)

a. Close manifold valves *a* and *b*.
b. Connect color coded hoses as shown.
c. Simultaneously close the low side service valve *c* while stopping the compressor.
d. Read low side pressure.
e. Wait several minutes.
f. Reread pressure. If valves are good reading will not change appreciably.

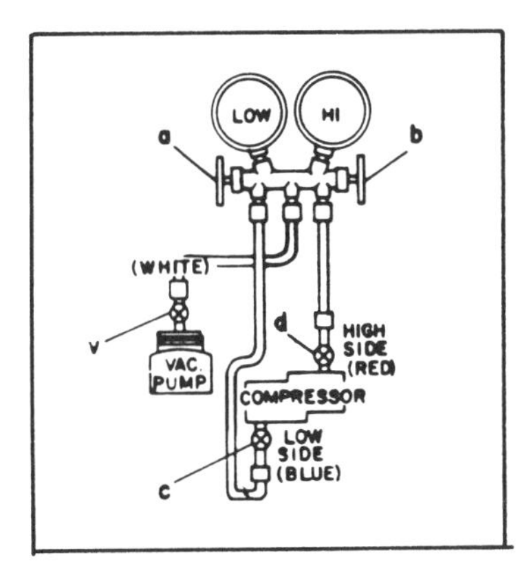

FIGURE 3.8

8. Low Side (Fig. 8)

a. Be sure system is not operating.
b. Close manifold valves *a* and *b*.
c. Purge all refrigerant from system and/or compressor.
d. Connect color coded hoses as shown.
e. Front seat service valves *c* and *d*.
f. Open manifold valve *a* and vacuum pump valve *v*. Start vacuum pump and run until system stabilizes.
g. Close manifold valve *a* and vacuum pump valve *v* and turn off vacuum pump.
h. Read and record vacuum.
i. Wait several minutes.
j. Reread low side compound gauge. If valves are good, readings will not change appreciably.

Courtesy of Copper Development Association

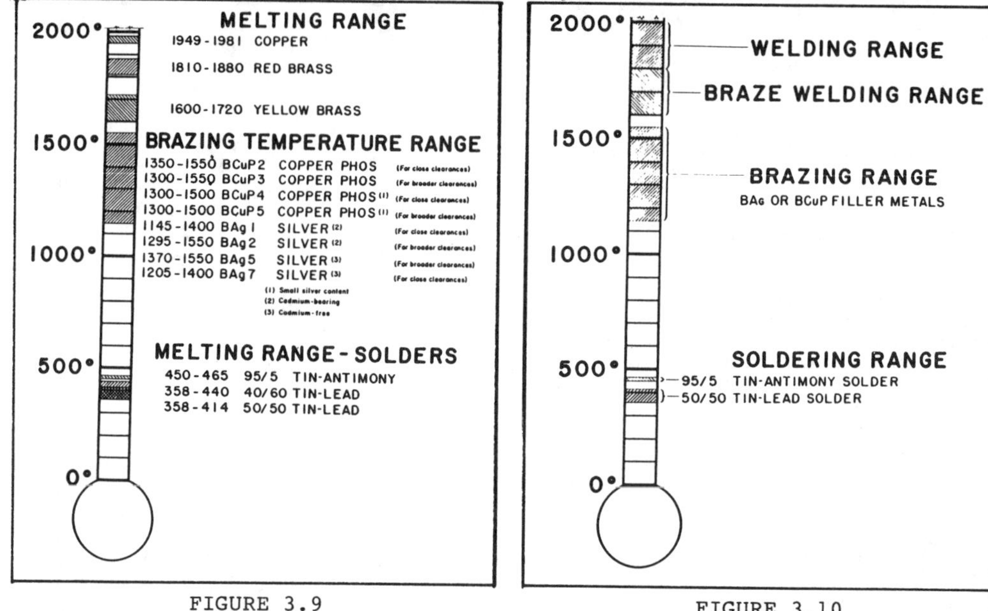

FIGURE 3.9

WELDING, BRAZING AND SOLDERING RANGES

FIGURE 3.10

BEHAVIOR OF FLUX IN BRAZING CYCLE

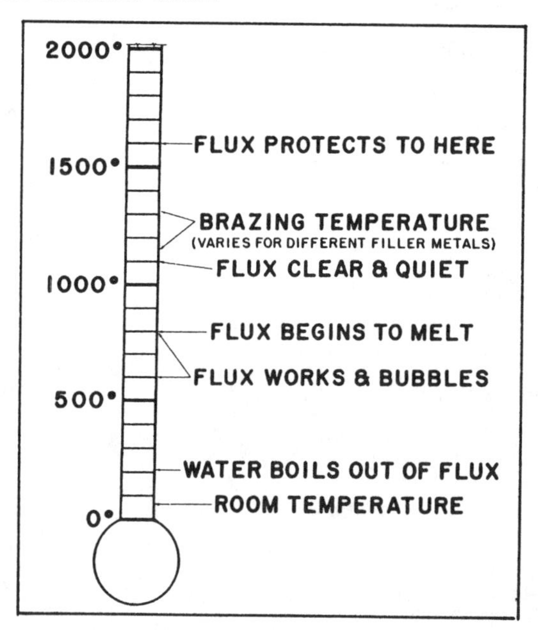

FIGURE 3.11

MELTING AND BRAZING RANGES

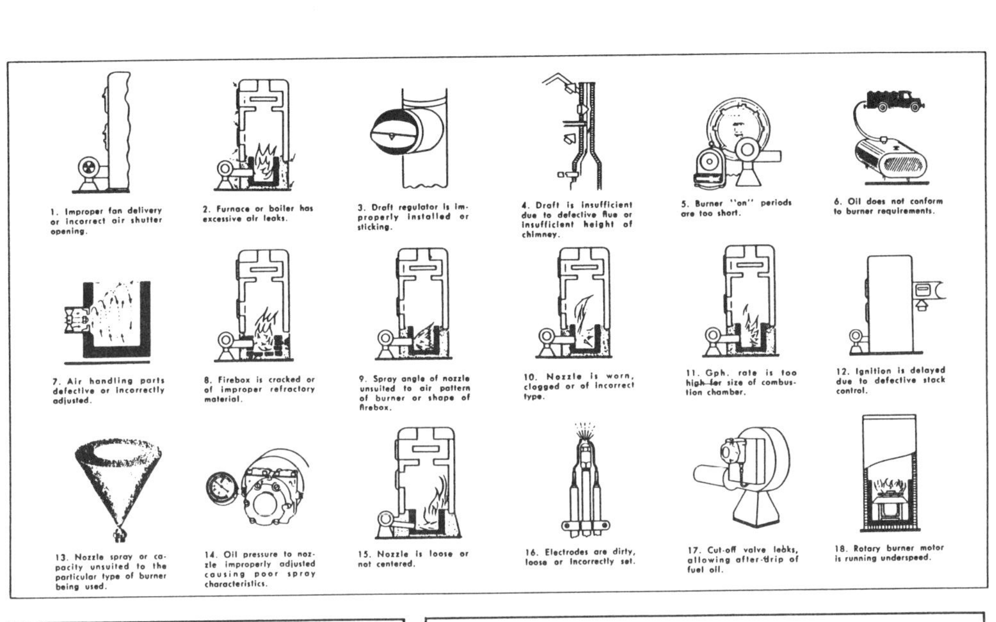

CO_2, EXCESS AIR AND STACK LOSS

% CO_2	% Excess Air	Net Stack Temperature Degrees Fahrenheit 400	500	600	700	800
15	2	14.75	16.75	19.0	21.25	23.5
14	7	15.25	17.25	19.75	22.0	24.5
13	15	15.75	18.0	20.5	23.0	25.5
12	25	16.25	18.75	21.5	24.25	27.0
11	35	17.0	19.75	22.75	25.5	28.5
10	50	18.0	21.25	24.25	27.25	30.5
9	65	19.25	22.75	26.0	29.25	33.0
8	78	20.75	24.25	28.25	32.0	36.0
7	90	22.75	26.75	31.0	35.25	39.75
6	110	25.0	30.0	34.75	30.75	45.0
5	130	28.25	34.25	40.0	46.0	52.0

SAVINGS FOR EVERY $100 FUEL COSTS BY INCREASE OF COMBUSTION EFFICIENCY

ASSUMING CONSTANT RADIATION AND OTHER UNACCOUNTED FOR LOSSES

From an Original Efficiency of:	To an Increased Combustion Efficiency of: 55 %	60 %	65 %	70 %	75 %	80 %	85 %	90 %	95 %
50 %	$9.10	$16.70	$23.10	$28.60	$33.30	$37.50	$41.20	$44.40	$47.40
55 %	—	8.30	15.40	21.50	26.70	31.20	35.30	38.90	42.10
60 %	—	—	7.70	14.30	20.00	25.00	29.40	33.30	37.80
65 %	—	—	—	7.10	13.30	18.80	23.50	27.80	31.60
70 %	—	—	—	—	6.70	12.50	17.60	22.20	26.30
75 %	—	—	—	—	—	6.30	11.80	16.70	21.10
80 %	—	—	—	—	—	—	5.90	11.10	15.80
85 %	—	—	—	—	—	—	—	5.60	10.50
90 %	—	—	—	—	—	—	—	—	5.30

FIGURE 3.12

COMMON CAUSES OF LOW CO_2 AND SMOKY FIRE ON OIL BURNERS

Courtesy The Bacharach Company

COOLING TOWERS

A cooling tower is simply an air-to-water-to-refrigerant heat exchanger. An air-cooled condenser is an air-to-refrigerant heat exchanger. In both cases the heat accumulated in the condenser is given up to the atmosphere.

A water cooling tower is more efficient than an air-cooled condenser because water can be cooled to a point below the surrounding ai by evaporative cooling. The theoretical cooling temperature is wet-bulb + 5°F. Miami design conditions are 91°F db and 79°F wb; therefore, the water could be cooled to 79 + 5 = 84°F, whereas the air-cooled system would operate at 91-95°F.

Figure 3.3 shows a typical atmospheric cooling tower. The static head in Figure 3.13 , is 7 feet. The difference between the hot condenser water discharge and the cooled water temperature leaving the tower is called the RANGE The difference in temperature between the cooled water temperature and the ambient wet-bulb is called the APPROACH.

To calculate tower water circulation for average comfort cooling application, an acceptable rule of thumb is 3 gpm per ton of cooling.

Actual calculations require these formulas:

1. Gpm = Btuh/500 x range
2. Btuh = gpm x 500 x range
3. Range = Btuh/500 x gpm

To solve for pump horsepower for cooling tower, condenser water or chilled water;

$$\text{Bhp} = \frac{\text{gpm} \times \text{ft head}}{3960 \times \text{eff}}$$

If the efficiency is not known assume 70%, or

$$\text{Bhp} = \frac{\text{gpm} \times \text{ft head}}{3960 \times .70}$$

Range = Hot Water Temp - Cold Water Temp

Approach = Cold Water Temp - Ambient WB Temp

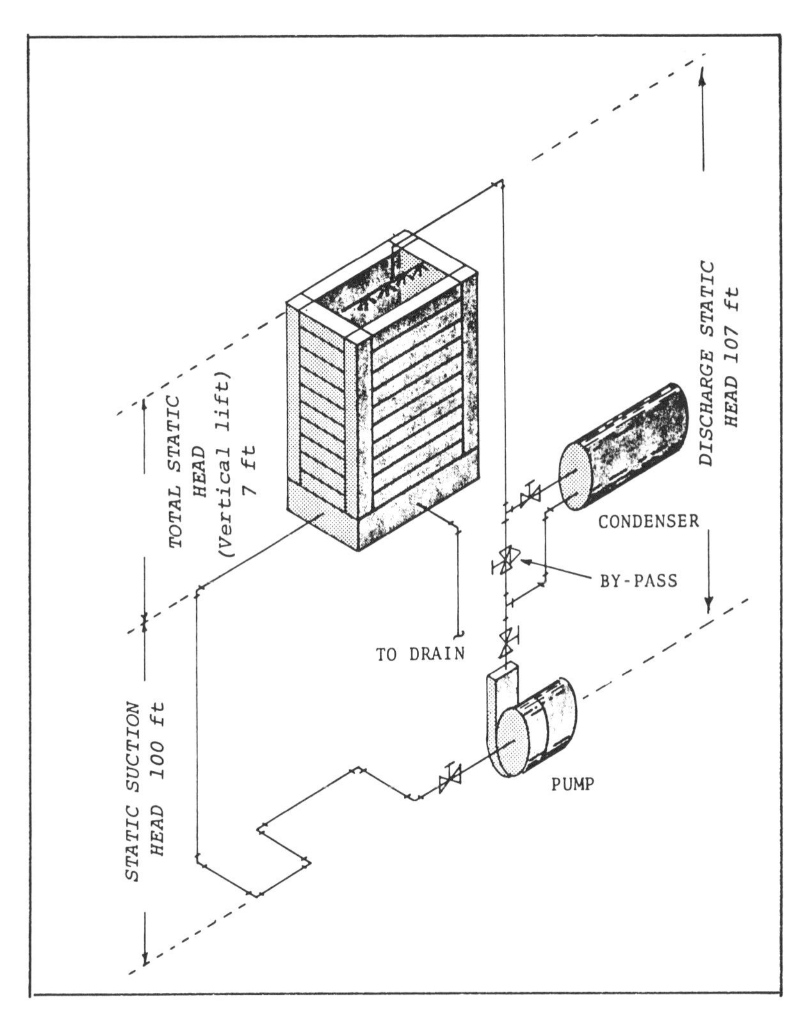

FIGURE 3.13

Total Static Head = Static Discharge - Static Suction

Note: Not shown is the bleed-off or blowdown line. The purpose of such a line is to provide a flushing action to remove some of the concentrations of solids normally occurring in the system. The bleed-off should be connected at the leaving side of the condenser and run into a suitable drain or sewer. The float in the tower sump will allow for an equal ammount of make-up water to enter the system.

OFFSETS

When an obstruction occurs in the way of a pipe run, the fitter must "break" around the obstruction to form an *offset*, thereby continuing the pipe run in the same direction but no longer in the same line. Often, the pipefitter can "eyeball" the job. With the help of a rule, straight-edge or string he can achieve a neater and closer fitting than merely eyeballing.

But there are times where the break is critical and the fitter must resort to calculations. Almost every examination will have questions dealing with pipe fitting problems. Figure 3.14 shows a standard 45° offset bend with the usual terminology. Sometimes the *run* is called the "advance" and some books refer to the *travel* as the "longside". Table 3.4 gives the solutions for offset calculations based on the nomenclature used in Figure 3.14.

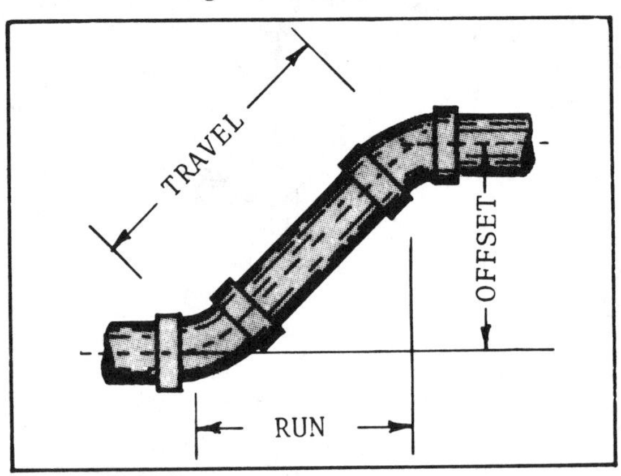

FIGURE 3.14

OFFSET BENDS

The formulas for several offset bends are given below. Most examinations will have some question on a double offset 540° expansion bend or a 360° expansion U-bend. The calculations of offset bends are always a function of the radius; the radius will usually be given and you will be asked to find the other dimensions.

TABLE 3.4

FORMULAS FOR OFFSET CALCULATIONS

To Find	Multiply	For 45° Ells	For 22-1/2° Ells	For 11-1/2° Ells	For 5-5/8° Ells	For 60° Ells	For 30° Ells	For Angle Bends
Travel	Offset x	1.414	2.613	5.126	10.207	1.155	2.0	Cosec
Offset	Travel x	0.707	0.383	0.195	0.098	0.866	0.5	Sin
Run	Offset x	1.0	2.414	5.027	10.158	0.577	1.732	Cot
Offset	Run x	1.0	0.414	0.199	0.098	1.732	0.577	Tan
Travel	Run x	1.414	1.082	1.019	1.005	2.00	1.155	Sec
Run	Travel x	0.707	0.933	0.981	0.995	0.500	0.866	Cos

See Table 3.5 for the calculated developed lengths of various bends and radii.

1. 240° Crossover bend

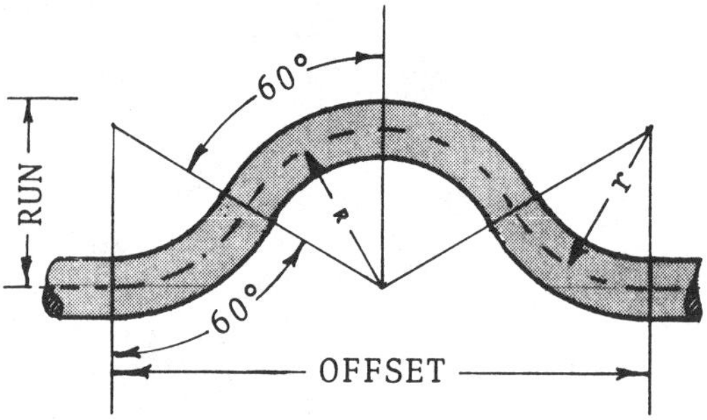

FIGURE 3.15

Offset = Radius
Run = Radius x 3.464
Developed length of pipe = Radius x 4.189

2. 360° Expansion U-bend

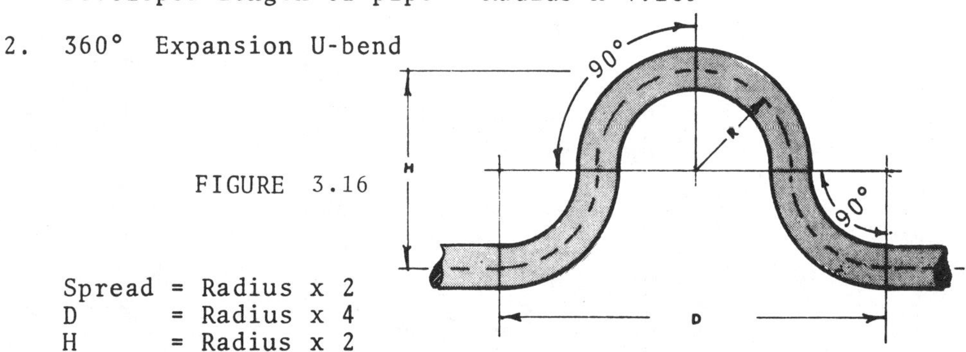

FIGURE 3.16

Spread = Radius x 2
D = Radius x 4
H = Radius x 2
Developed length of pipe = Radius x 6.283

3. 540° Double offset expansion bend

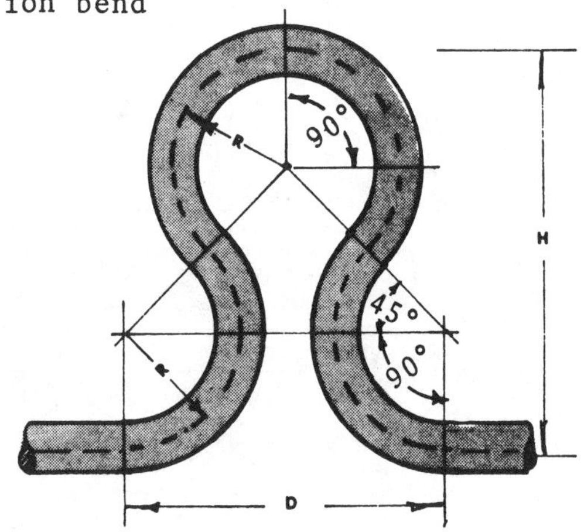

FIGURE 3.17

Spread = Radius x 0.8284
D = Radius x 2.828
H = Radius x 3.414
Developed length of pipe = Radius x 9.425

4. 180° U-bend

FIGURE 3.18

Spread = Radius x 2
H = Radius
Developed length of pipe = Radius x 3.142

5. 360° Circle bend

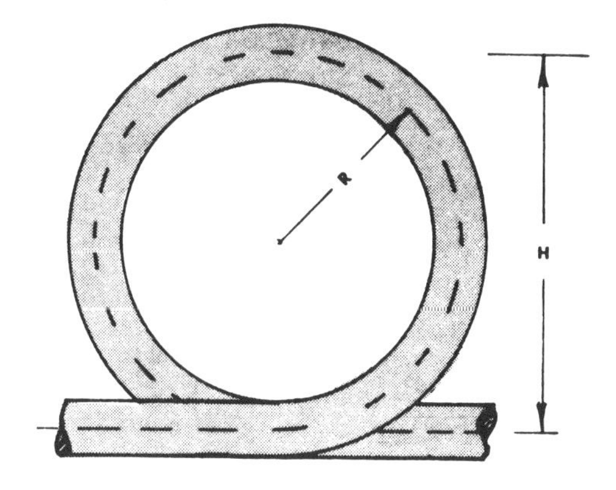

FIGURE 3.19

Spread = Radius x 2
H = Radius x 2
Developed length of pipe = Radius x 6.283

6. 3 Piece 90° turn

In the layout for fabrication of a 3 piece 90° turn there are two mitered joints. Angles of cuts and pipe lengths are figured as follows:

$$\text{Angle of cut} = \frac{\text{number of degrees of turn}}{2 \times \text{number of welds}} \quad \text{or,}$$

$$\frac{90}{2x2} = \frac{90}{4} = 22\text{-}1/2°$$

a = Radius x 0.414
A = 2a = Radius x 0.828
Developed length = 3 A
Cut back distance =
pipe O.D. x 0.414 =

Example: From Figure 3.20 assuming 4" schedule 40 pipe and radius of 12 in.;

a = 12 in. x 0.414	=	4.97 in.
A = 2 x 4.97	=	9.94 in.
Developed length - 3 x 9.94	=	29.82 in.
Cut back = 4.5 x 0.414	=	1.86 in.
B = 9.94 + 1.86	=	11.80 in.
C = 9.94 - 1.86	=	8.08 in.

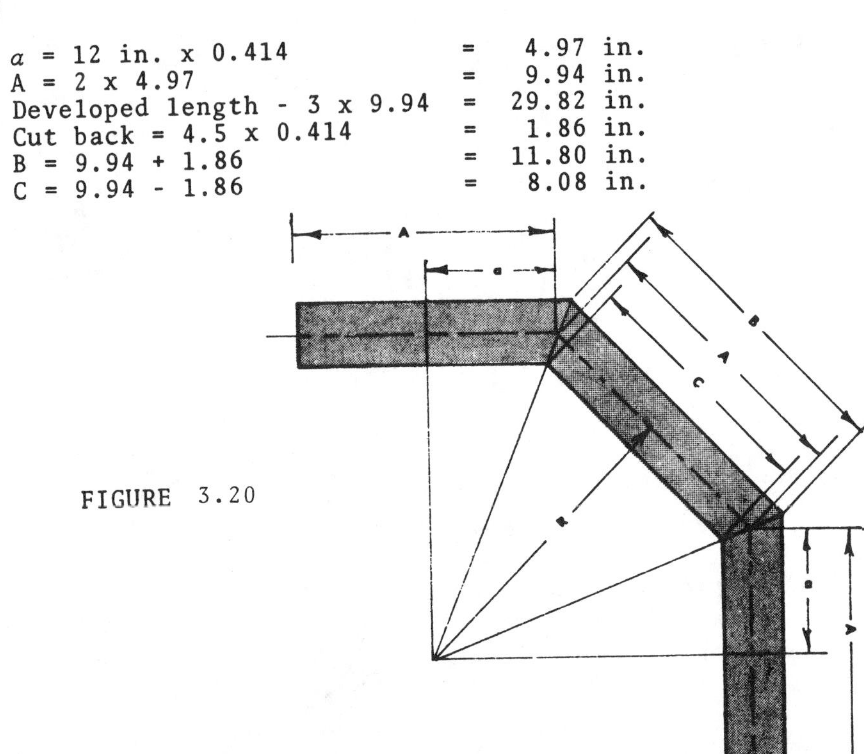

FIGURE 3.20

7. Three run 45° equal spread

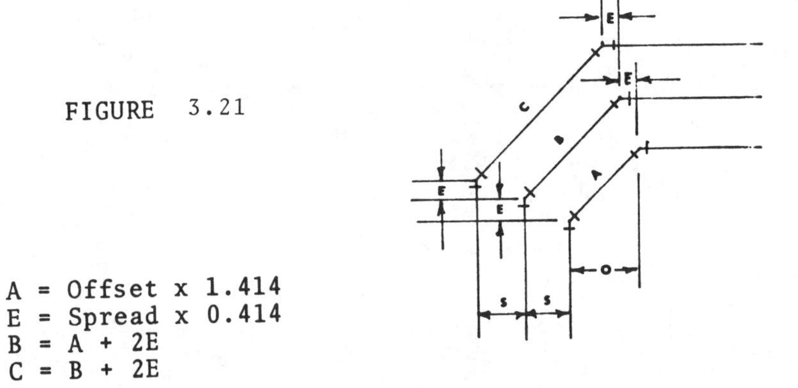

FIGURE 3.21

A = Offset x 1.414
E = Spread x 0.414
B = A + 2E
C = B + 2E

TABLE 3.5

CALCULATED LENGTHS OF PIPE FOR VARIOUS BENDS

INCHES	FEET	90°	180°	270°	360°	540°
1		1½"	3"	4¾"	6¼"	9½"
2		3	6¼	9½	12½	18¾
3	¼	4¾	9½	14¼	18¾	28¼
4		6¼	12½	18¾	25¼	37¾
5		7¾	15¾	23½	31½	47¼
6	½	9½	18¾	28¼	37¾	56½
7		11	22	33	44	66
8		12½	25¼	37¾	50¼	75½
9	¾	14¼	28¼	42½	56½	84¾
10		15¾	31½	47¼	62¾	94¼
11		17¼	34½	51¾	69	103¾
12	1	18¾	37¾	56½	75½	113
12.5		19¾	39¼	59	78½	117¾
14		22	44	66	88	132
15	1¼	23½	47	70¾	94¼	141¼
16		25¼	50¼	75½	100½	150¾
17.5		27½	55	82½	110	165
18	1½	28¼	56½	84¾	113	169¾
20		31½	62¾	94¼	125¾	188½
21	1¾	33	66	99	132	198
24	2	37¾	75½	113	150¾	226¼
25		39¼	78½	117¾	157	235½
30	2½	47¼	94¼	141¼	188½	282¾
32		50¼	100½	150¾	201	301½
36	3	56½	113	169½	226¼	339¼
40		62¾	125¾	188½	251¼	377
48	4	75½	150¾	226¼	301½	452½
50		78½	157	235½	314¼	471¼
56		88	176	264	351¾	527¾
60	5	94¼	188½	282¾	377	565½
64		100½	201	301½	402	603¼
70		110	220	329¾	439¾	659¾
72	6	113	226¼	339¼	452½	678½
80		125¾	251¼	377	502¾	754
84	7	132	263¾	395¾	527¾	791½
90	7½	141¼	282¾	424	565½	848¼
96	8	150¾	301½	452½	603	904¾
100		157	314¼	471¼	628¼	942½
108	9	169½	339¼	509	678½	1017¾
120	10	188½	377	565½	754	1131
132	11	207¼	414¾	622	829½	1244
144	12	226¼	452½	678½	904¾	1357¼
156	13	245	490	735¼	980¼	1470¼
168	14	263¾	527¾	791½	1055½	1583½
180	15	282¾	565½	848¼	1131	1696½
192	16	301½	603	904¾	1206¼	1809½
204	17	320½	640¾	961¼	1281¾	1922½
216	18	339¼	678½	1017¾	1357¼	2035¾
228	19	358	716¼	1074½	1432½	2148¾
240	20	377	754	1131	1508	2262

To find the length of pipe in a bend having a radius not given above, add together the length of pipe in bends whose combined radii equal the required radius.

Example: **Find length of pipe in 90° bend of 5′ 9″ radius.**

Length of pipe in 90° bend of 5′ radius = 94¼″
Length of pipe in 90° bend of 9″ radius = 14¼″

Then, length of pipe in 90° bend of 5′ 9″ radius = 108½″

PITCHING PIPE

To Find the Drop in a Pipe Run: multiply the pitch per foot X the length of run in feet.

What is the drop for a 100 ft run of pipe with a 1/4 in. pitch?

100 x 1/4 in. = 100 x .25 = 25 in., 25 ÷ 12 in. = 2.08 ft.

To Find the Pitch in a Run of Pipe: divide the drop (inches) by the length of run (feet).

What is the pitch for a 100 ft run of pipe with a 25 in. drop?

$$\frac{25 \text{ in.}}{100 \text{ ft}} = .25 \text{ inch} = 1/4 \text{ in. pitch}$$

EXPANSION OF PIPE

To calculate the expansion and contraction of pipe it is necessary to know the *coefficient of expansion* for the particular material. Table 3.6 lists various coefficients.

TABLE 3.6

COEFFICIENT OF EXPANSION OF PIPE

METAL	COEFFICIENT OF EXPANSION PER °F
Cast Iron	0.0000059
Copper	0.0000094
Steel	0.0000061
Wrought Iron	0.0000069
Brass	0.0000104

Once the coefficient of expansion is known the formula can be applied;

$$E = C \times L \times \Delta T$$

where

E = Expansion of pipe (inches)
C = Coefficient of expansion
L = Length of run (inches)
ΔT = Temperature change

When the temperature change is not exactly known, it is customary to use a 100°F rise for standard cooling and heating work.

The movement of a 3 inch copper hot-water pipe, 100 ft long when the temperature changes from 70°F to 170°F is equal to;

0.0000094 x 100°F x 100 ft x 12 in. = 1.128 inches

Sizing the Expansion Loop: Any pipe subject to temperature change will expand and contract with each change in a lengthwise direction, in accordance with the above formula. It is therefore essential to provide an expansion section. Table 3.7 is taken from Mueller Brass Co. data; it lists the radii required for different sizes of copper tube to take up to 6 inches of expansion.

TABLE 3.7

Pipe Size Inches	Radius in Inches (R) For Expansion Of								
	1/2"	1"	1-1/2"	2"	2-1/2"	3"	4"	5"	6"
3/4	10	15	19	22	25	27	30	34	38
1	11	16	20	24	27	29	33	38	42
1-1/4	11	17	21	26	29	32	36	42	47
1-1/2	12	18	23	28	31	35	39	46	51
2	14	20	25	31	34	38	44	51	57
2-1/2	16	22	27	32	37	42	47	56	62
3	18	24	30	34	39	45	53	60	67
4	20	28	34	39	44	48	58	66	75
5	22	31	39	44	49	54	62	70	78
6	24	34	42	48	54	59	68	76	83

Bends to the left of the stepped line may be made from 20 ft lengths or less. Bends requiring more than 20 ft lengths must be made up in sections and assembled with couplings or flanges. The expansion section should be cold sprung approximately one-half of the expected distance of expansion.

of expansion

After the number of inches of expansion has been determined from the above formula, the required size expansion loop may be selected from Table 3.7. Still using the above example (the expansion was calculated at 1.128 inches) interpolating from Table 3.5, the required radius for 3 in. pipe is 26 inches. Figures 3.22 and 3.23 show the pipe shape and length for expansion bends and offsets.

OFFSET WITH FOUR 90° CAST ELLS

FIGURE 3.22

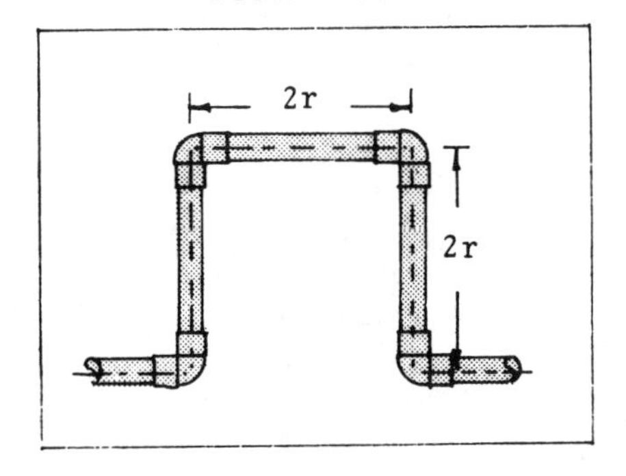

EXPANSION LOOP

FIGURE 3.23

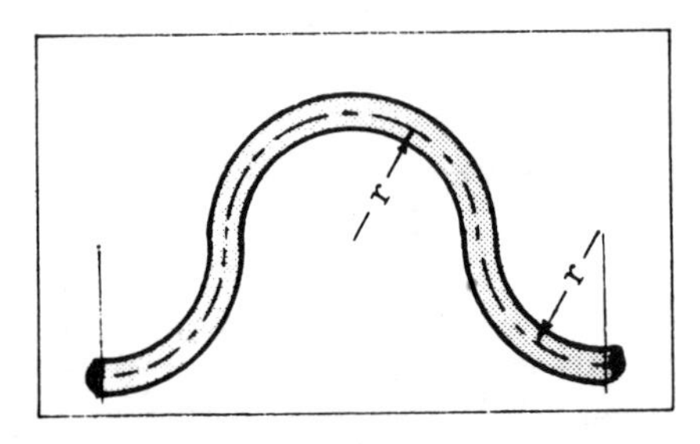

Thus, from Figure 3.22, 13.0 ft of pipe will be required and from Figure 3.23, 13.5 ft of pipe will be required.

If the example called for steel pipe rather than copper the calculations would be somewhat different. Using the Formula $E = C \times L \times \Delta T$;

$$0.0000061 \times 100°F \times 100 \text{ ft} \times 12 \text{ in.} = 0.732 \text{ in. expansion}$$

For steel pipe to 20 inches, and expansion of up to 10 inches, the expansion loop may be calculated from Figure 3.12

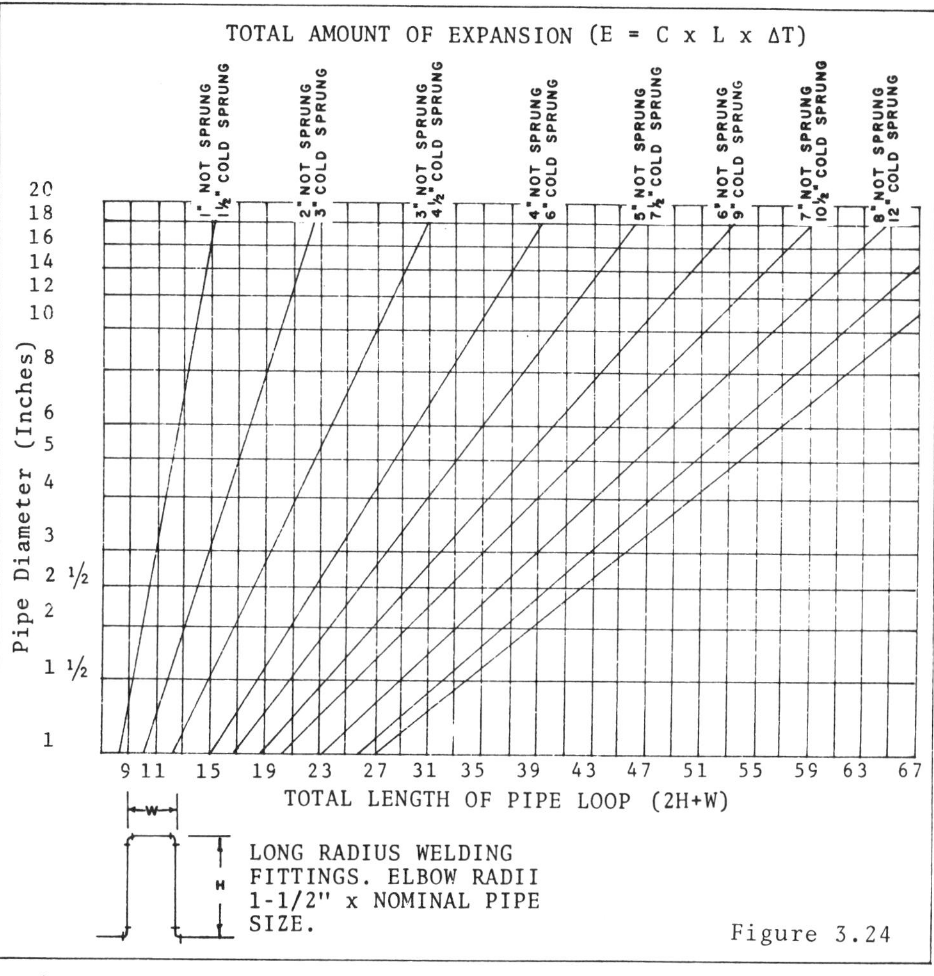

Figure 3.24

Given: A 100 ft run of 3 in. steel pipe at 100° F rise.

Find:
1. Expansion in inches
2. Size of expansion loop

Step 1. From Table 3.4, the coefficient of expansion for steel pipe is 0.0000061

Step 2. Substituting for E = CxLxΔT;
0.0000061 x 100°F x 100 ft x 12 in. = 0.732 in. expansion

Step 3. Enter the left ordinate of the chart in Figure 3.24 at 3 in. pipe diameter. Follow the horizontal line to the curve for 1" not sprung. Extrapolating the curve, the required length of the loop is 10 ft. Maintaining the recommended ratio;

$W = \frac{H}{2} = \frac{4}{2}$, therefore, W=2', H=4', H=4'.

FIGURE 3.25

HOW TO READ REDUCING FITTINGS

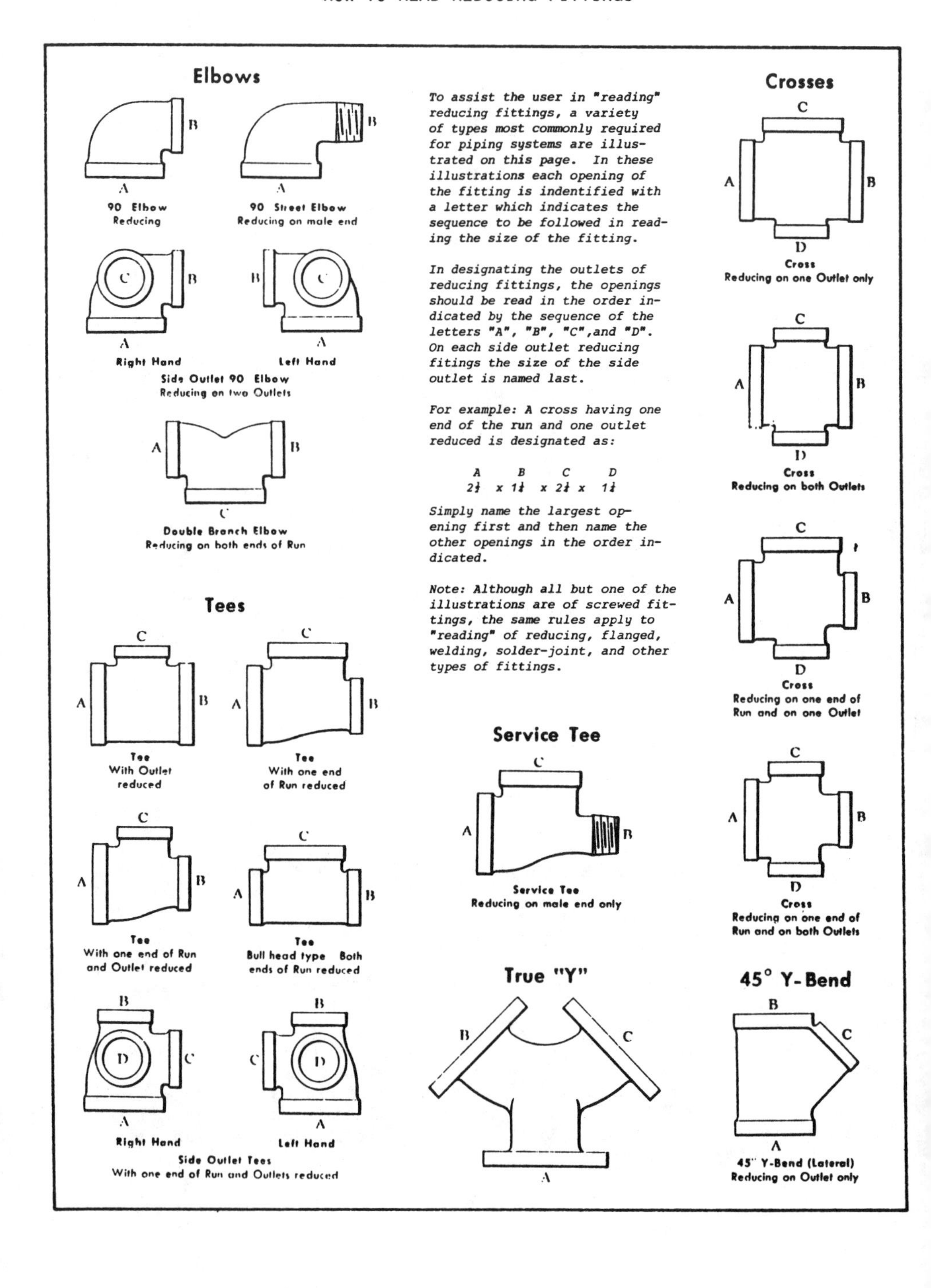

FIGURE 3.26

HOW TO IDENTIFY FLANGE FITTINGS

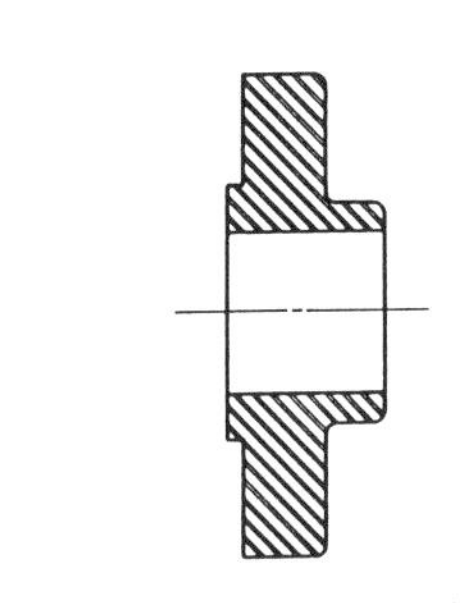

(a) Raised-face flange. Applicable to pressures to 900 lb.

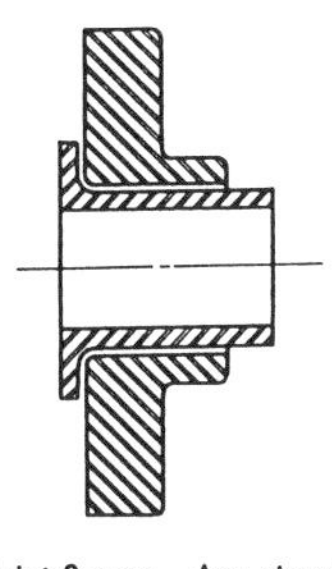

(b) Lap-joint flange. Any standard facing may be machined into upset face.

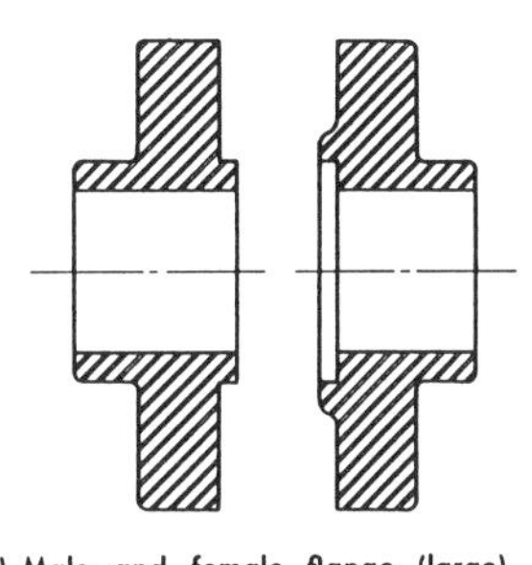

(c) Male and female flange (large). Use when desirable to have a facing that retains the gasket.

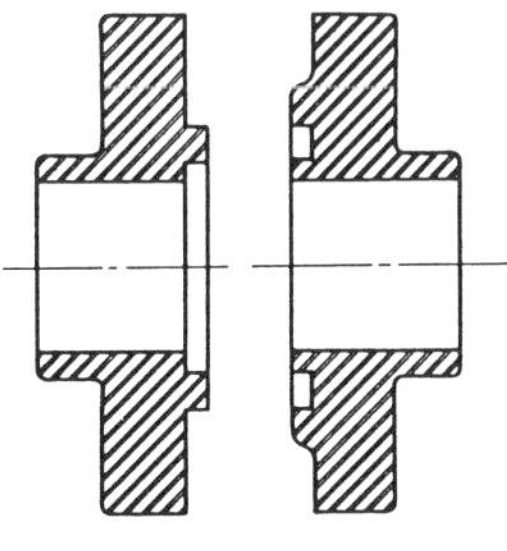

(d) Tongue and groove flange (large). Use when desirable to have a facing that retains the gasket.

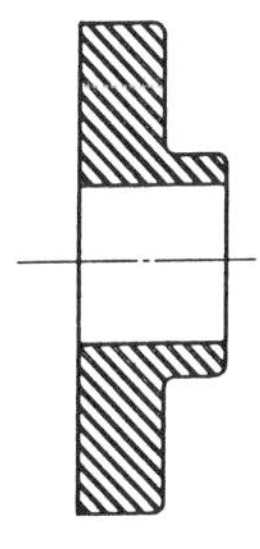

(e) Flat-face flange. Primarily used on cast iron.

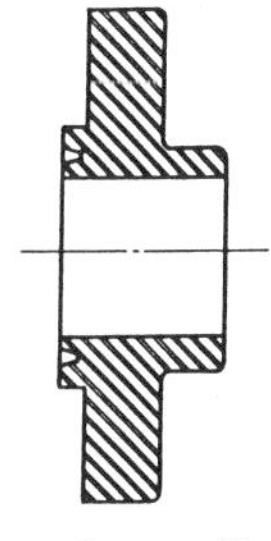

(f) Ring-joint flange. Use for high-pressure applications.

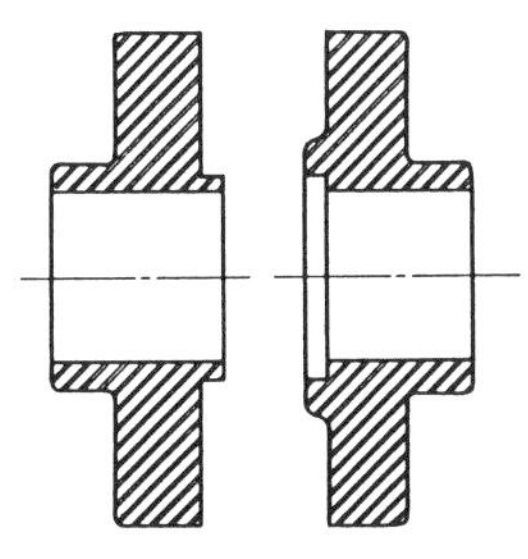

(g) Male and female flange (small). Same use as large male and female flange, except it provides for higher gasket compression.

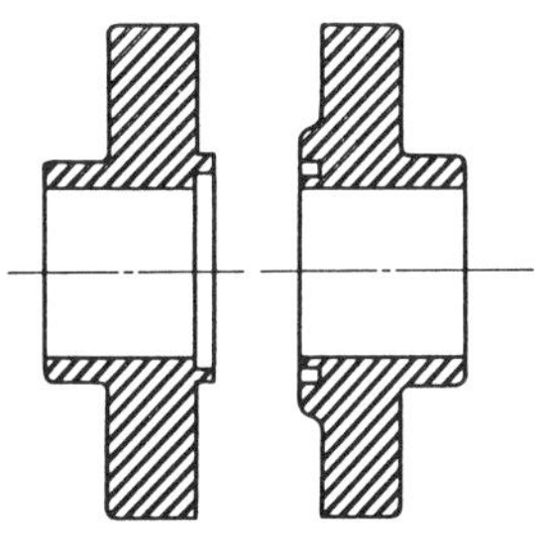

(h) Tongue and groove flange (small). Same use as large tongue and groove flange, except it provides for higher gasket compression.

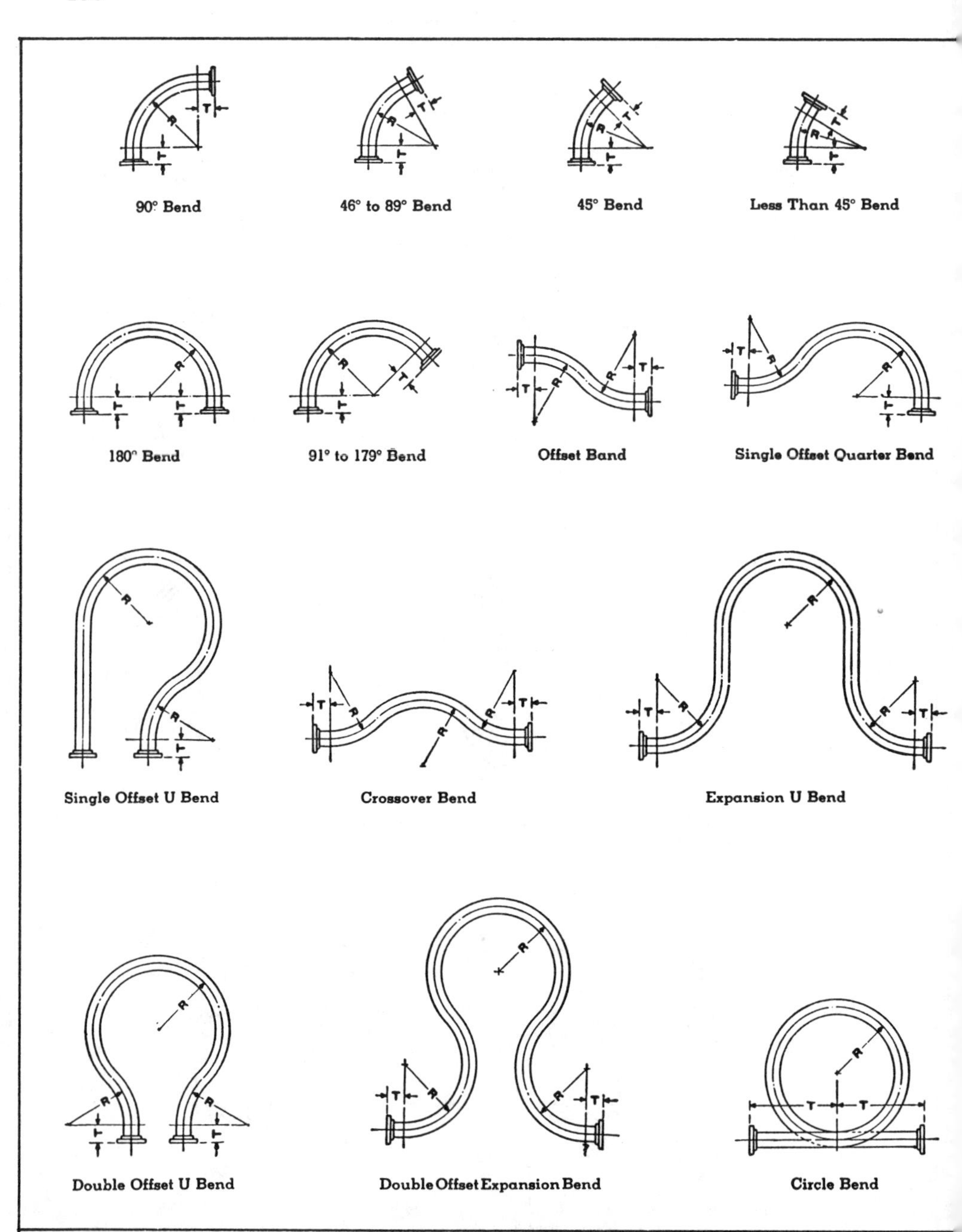

FIGURE 3.27

HOW TO IDENTIFY PIPE BENDS

TABLE 3.8

HANGER AND SUPPORTS

To find the recommended hanger or support components, locate the system temperature and insulation conditions in the two columns at the left. Then read across the column headings for the type component to be used. Numbers in the boxes refer to those types shown in Figure 3.28.

SYSTEM	INSULATION	HORIZONTAL PIPE ATTACHMENTS										VERTICAL PIPE ATTACHMENTS			HANGER ROD FIXTURES			BUILDING STRUCTURE ATTACHMENTS				
															Steel or Mall. Iron			Steel and/or Mall. Iron				
Temp. Range deg. F.	Note 1	Steel Clips A	Mall. Iron Rings B	Steel Bands C	Steel Clamps D	Cast Iron Hanging Rolls E	Cast Iron Supporting Rolls F	Steel Trapezes G	Steel Prot. Saddles & Shields H	Steel or Cast Iron Stanchions I	Steel Welded Attachments J	Steel Riser Clamps 2-Bolt K	Steel Riser Clamps 4-Bolt L	Welded Attachments Steel M	Turn-Buckles N	Swing Eyes O	Clevises P	Inserts Q	C-Clamps R	Beam Clamps S	Welded Attachments T	Brackets U
Hot A-1	Covered	None	None	1,7,9,10 w/Saddle or Shield	3	41,42,43 w/ Saddle	44,45,46,47 w/ Saddle	Note 3 w/ Saddle	39A,39B	35,36,37,38 w/ Saddle	Note 3	8	Note 3	Note 3	13,15	16,17	14	18,19 Note 5	23	20,21,25, 27,28,29,30	22 or Note 3	31,32,33,34
120° to 450°	Bare	24,26	5,6,11,12	1,7,9,10	3,4	41,42,43	44,45,46,47	Note 3	None	35,36,37,38												
Hot A-2	Covered	None	None	1 w/ Saddle	3	41,42 w/ Saddle	41,42 w/ Saddle	Note 3 w/ Saddle	39A,39B	35,36,37,38 w/ Saddle	Note 3	Note 3	Note 3	Note 3	13,15	16,17	14	18,19 Note 5	23	20,21,25 27,28,29,30	22 or Note 3	31,32,33,34
451° to 750°	Bare	None	None	None	3,4	None	None	None	None	None												
Hot A-3	Covered	None	None	1 w/ Alloy Saddle	Alloy 2,3	41,42 w/ Alloy Saddle	41,42 w/ Alloy Saddle	Note 3 w/ Alloy Saddle	39A,39B Alloy	35,36,37,38 w/ Alloy Saddle	Note 3 Alloy	Note 3	Alloy Note 3	Alloy Note 3	13	17	14	Note 3	None	20,21,25 27,28,29,30	22 or Note 3	31,32,33,34
Over 750°	Bare	None	None	None	Alloy 2,3,4	41,42 w/ Alloy Saddle	41,42 w/ Alloy Saddle	None	None	None												
Ambient B	Covered	24,26	None	1,7,9,10 w/Saddle or Shield	3	41,42,43 w/ Saddle	44,45,46,47 w/ Saddle	Note 3 w/ Saddle	39A,39B,40	35,36,37,38 w/ Saddle	Note 3	8	Note 3	Note 3	13,15	16,17	14	18,19 Note 5	23	20,21,25 27,28,29,30	22 or Note 3	31,32,33,34
60° to 119°	Bare	24,26	5,6,11,12	1,7,9,10	3,4	41,42,43	44,45,46,47	Note 3	None	35,36,37,38												
Cold C-1	Covered	None	None	1,7,9,10 w/ Shield	3	41,42,43 w/ Shield Note 4	44,45,46,47 w/ Shield Note 4	Notes 3 & 4 w/ Shield	40	35,36,37,38 w/ Shield	Note 3	8	Note 3	Note 3	13,15	16,17	14	18,19 Note 5	23	20,21,25 27,28,29,30	22 or Note 3	31,32,33,34
33° to 59°	Bare	24,26	5,6,11,12	1,7,9,10	3,4	41,42,43	44,45,46,47	Note 3	None	35,36,37,38												
Cold C-2	Covered	None	None	1,7,9,10 w/ Shield	None	41,42,43 w/ Shield Note 4	44,45,46,47 w/ Shield Note 4	Notes 3 & 4 w/ Shield	40	35,36,37,38 w/ Shield	Note 3	8	Note 3	Note 3	13,15	16,17	14	18,19 Note 5	23	20,21,25 27,28,29,30	22 or Note 3	31,32,33,34
-20° to 32°	Bare	None	None	1,7,9,10	3,4	41,42,43	44,45,46,47	Note 3	None	35,36,37,38												
Cold C-3	Covered	None	None	1,7,9,10 w/ Shield	None	41,42,43 w/ Shield Note 4	44,45,46,47 w/ Shield Note 4	Notes 3 & 4 w/ Shield	40	35,36,37,38 w/ Shield	Notes 3 & 4	Note 2	Note 2	Notes 2 & 3	13,15	16,17	14	18,19 Note 5	23	20,21,25 27,28,29,30	22 or Note 3	31,32,33,34
Below -20°	Bare	None	None	Note 2	Note 2	Note 2	Note 2	Note 2	None	Note 2												

1. Hangers on insulated systems shall incorporate protection saddles or shields or shall be clamped or welded immovably to the pipe and project through the cover to provide a point of attachment external to the insulation. 2. The selection of type and material shall be made by the engineering design. 3. The design shall be in accordance with MSS SP-58 or as specified by engineering design. 4. For shields used with rollers or subject to point loading, see table 4. 5. Continuous inserts, anchor bolts and expansion cases may be used as specified in the engineering design.

Courtesy of the Manufacturer's Standardization Society of the Valve and Fittings Industry.

TABLE 3.9

Minimum Sizes of Straps, Rods and Chains for Hangers
for Refrigeration Piping

Nominal Pipe Size (inches)	Component (Steel)	Minimum Stock Size (inches)	
		Exposed to Weather	Protected from Weather
1 & Smaller	Strap	1/8 Thick	1/16 Thick x 3/4 Wide
Above 1	Strap	1/4 Thick	1/8 Thick x 1 Wide
2 & Smaller	Rod	3/8 Diameter	3/8 Diameter
Above 2	Rod	1/2 Diameter	1/2 Diameter
2 & Smaller	Chain	3/16 Diameter or Equivalent Area	3/16 Diameter or Equivalent Area
Above 2	Chain	3/8 Diameter or Equivalent Area	3/8 Diameter or Equivalent Area
All Sizes	Bolted Clamps	3/16 Thick; bolts 3/8 Diameter	3/16 Thick; bolts 3/8 Diameter

Note: For nonferrous materials, the minimum stock area shall be increased by the ratio of allowable stresses of steel to the allowable stress of the nonferrous material.

Extracted from *Refrigeration Piping Code* (ANSI B31.5-1966) section of the ANSI Standard Code for Pressure Piping with permission of the publisher, The American Society of Mechanical Engineers, 345 East 47th St., New York.

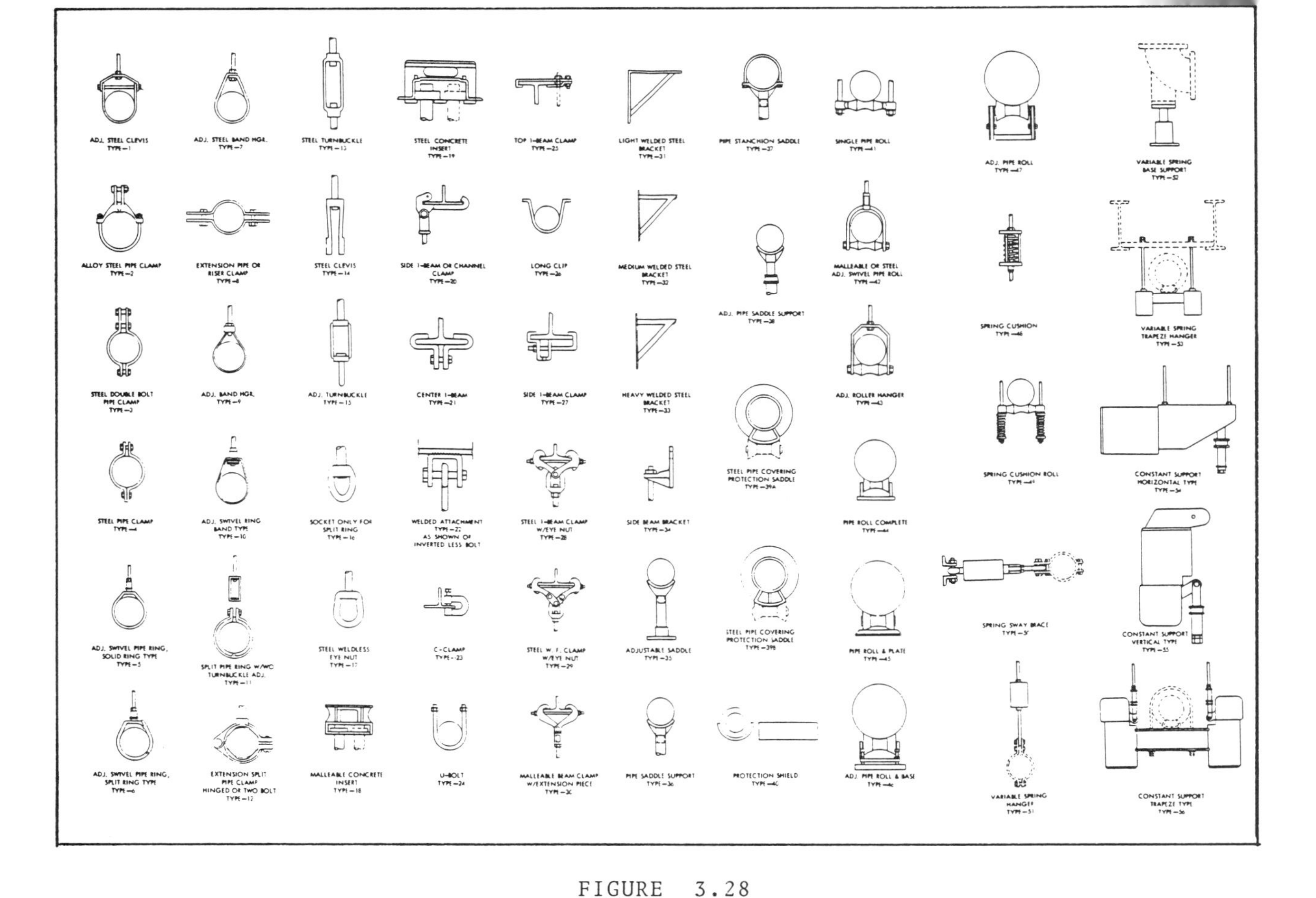

FIGURE 3.28

HANGER SELECTION

TABLE 3.10

LAYING PROJECTIONS FOR BUTT WELDING ELBOWS FOR STANDARD AND EXTRA STRONG STEEL PIPE.

This Table may be used to find the actual pipe length between fittings by subtracting the projections on each side from the calculated length.

	Dimension for *A* (inches)		
Nominal Diameter Inches	Long Radius 90° Elbow	Short Radius 90° Elbow	45° Elbow
1	1-1/2	1	7/8
1-1/4	1-7/8	1-1/4	1
1-1/2	2-1/4	1-1/2	1-1/8
2	3	2	1-3/8
2-1/2	3-3/4	2-1/2	1-3/4
3	4-1/2	3	2
3-1/2	5-1/4	3-1/2	2-1/4
4	6	4	2-1/2
5	7-1/2	5	3-1/8
6	9	6	3-3/4
8	12	8	5
10	15	10	6-1/4

TABLE 3.11

LAYING PROJECTIONS FOR CAST IRON SCREWED FITTINGS

ASA 125 Pound and 250 Pound Classes

Size	*Dimension, inches*							
	125-Pound						250-Pound	
	A	B	E	H	M Close	M Open	A	B
1/4	0.81	0.73	...	...	...	...	0.94	0.81
3/8	0.95	0.80	...	...	...		1.06	0.88
1/2	1.12	0.88	...	1.38	1.25	1.75	1.25	1.00
3/4	1.31	0.98	...	1.50	1.50	1.88	1.44	1.13
1	1.50	1.12	...	1.70	1.75	2.50	1.63	1.31
1 1/4	1.75	1.29	...	2.13	2.25	3.00	1.94	1.50
1 1/2	1.94	1.43	...	2.25	2.50	3.50	2.13	1.69
2	2.25	1.68	...	2.32	3.25	4.50	2.50	2.00
2 1/2	2.70	1.95	1.81	2.63	3.75	5.50	2.94	2.25
3	3.08	2.17	1.91	2.88	4.50	6.50	3.38	2.50
3 1/2	3.42	2.39	2.03	3.13	...	...	3.75	2.63
4	3.79	2.61	2.22	3.38	6.00	7.50	4.13	2.81
5	4.50	3.05	2.38	3.57	...	...	4.88	3.19
6	5.13	3.46	2.63	3.81	...	...	5.63	3.50
8	6.56	4.28	2.88	5.25	...	...	7.00	4.31
10	8.08	5.16	3.50	...	...	...	8.63	5.19
12	9.50	5.97	3.88	...	...	...	10.00	6.00

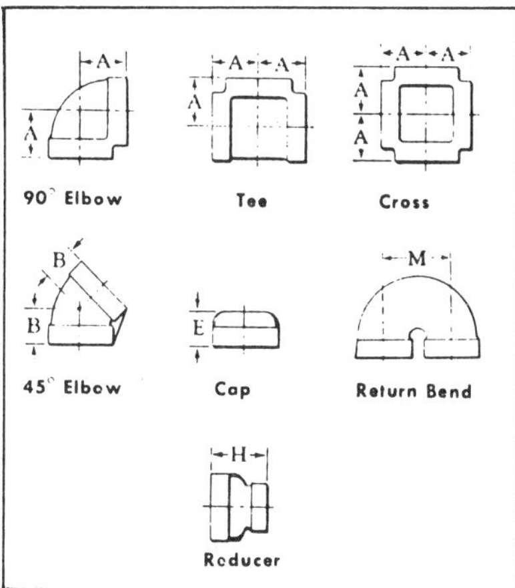

SOLVING SIMPLE PIPE MEASURING PROBLEMS

Example: Give the measurements for dimension "A" in Figure 3.34 for a 2 inch pipe LR butt weld 90 degree ells. Let "B" dimension remain as is.

Solution: From Table 3.10 find the dimension A for a 2 inch pipe = 3" per elbow x two elbows = 6". Then 16' 4-3/8" minus 6" = 15' - 10-3/8".

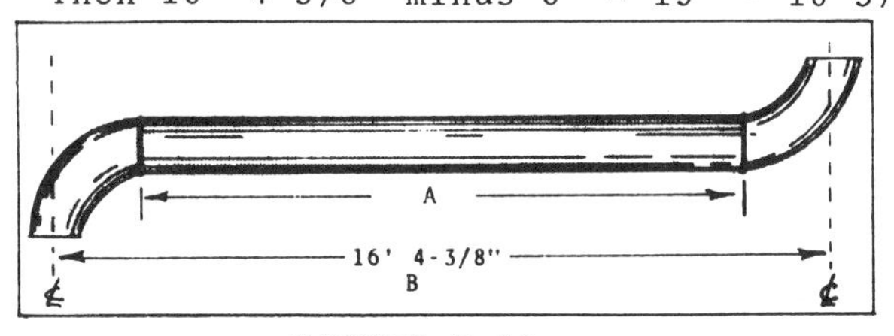

FIGURE 3.29

For screwed pipe, use Table 3.11. The easiest way to solve these kinds of problems is to convert the fractions to decimals and then perform the subtraction. Use Tables A.2 and A.3 in the Appendix to find decimals of a foot and inch-foot decimal conversions.

FANS, PULLEYS AND BELTS

System start-up, test and balance, and proper maintenance, require a thorough understanding of the mechanics and laws of pulleys and belts; and although fan law solutions can be worked out with great theoretical accuracy on paper, such fan laws are worthless unless a skilled journeyman can translate them into actual working conditions in the field. Testing and balancing of systems has only recently been receiving the attention it deserves , consequently a new awareness has been manifested and more questions are beginning to appear on *journeymens'* and *masters'* examinations.

FAN LAWS

The fan laws are an important tool for start-up, testing and balancing of any air-flow system. For a more thorough discussion of the fan laws see,AIR CONDITIONING/TESTING/ADJUSTING/ BALANCING, 2nd ed. by John Gladstone, Van Nostrand Rienhold,the TRANE AIR CONDITIONING MANUAL, the CARRIER HANDBOOK OF SYSTEM DESIGN, McGraw-Hill, and FAN ENGINEERING, Buffalo Forge Co.

The basic fans laws are:

1. $RPM_2 = RPM_1 \times \frac{CFM_2}{CFM_1}$

2. $CFM_2 = CFM_1 \times \frac{RPM_2}{RPM_1}$

3. $SP_2 = SP_1 \times \left(\frac{RPM_2}{RPM_1}\right)^2$

4. $HP_2 = HP_1 \times \left(\frac{RPM_2}{RPM_1}\right)^3$

With these formulas the technician can adjust fan speeds to meet new conditions.

Example:

A 1 hp fan is operating at 1" sp at a speed of 1000 rpm and developing 1200 cfm. If the speed is reduced to 800 rpm, what is the new cfm? the new sp? the new hp?

1. $CFM = 1200 \times \frac{800}{1000} = .80 \times 1200 = 960$ cfm

2. $SP = 1'' \times \left(\frac{800}{1000}\right)^2 = .80 \times .80 = 0.64$ in.

3. $HP = 1 \times \left(\frac{800}{1000}\right)^3 = .80 \times .80 \times .80 = 0.51$ hp

PULLEYS

Drive sets for fans and blowers consist of a driver pulley on the motor shaft a driven pulley on the blower shaft and a belt or set of matched belts to transmit the power. Pulley formulas are usually given in pulley diameters, for accuracy they should be considered in actual pitch diameters. Figures 3.30, 3.31, 3.32 and Table 3.11 give dimension for standdard variable sheaves.

The four basic pulley laws are:

1. $\text{rpm of driven} = \frac{\text{diameter of driver} \times \text{rpm}}{\text{diameter of driven}}$

2. $\text{rpm of driver} = \frac{\text{diameter of driven} \times \text{rpm}}{\text{diameter of driver}}$

3. $\text{Diameter of driven} = \frac{\text{diameter of driver} \times \text{rpm}}{\text{rpm of driven}}$

4. $\text{Diameter of the driver} = \frac{\text{diameter of driven} \times \text{rpm}}{\text{rpm of driver}}$

TABLE 3.11

VARIABLE SHEAVE GROOVE DIMENSIONS

Cross Section	b_g Closed (Inches)	b_g Open (Inches)	h_g Minimum (Inches)	2a (Inches)	$2a_v$ (Inches)	S_e Open Minimum (Inches)	S Minimum (Inches)
1430V	0.875 ± 0.005	1.582 ± 0.005	1.758	0.20	2.64	0.882	1.765
1930V	1.188 ± 0.005	2.142 ± 0.005	2.341	0.25	3.56	1.163	2.325
2530V	1.563 ± 0.007	2.823 ± 0.007	3.038	0.30	4.70	1.501	3.003
3230V	2.000 ± 0.007	3.665 ± 0.007	3.855	0.35	6.21	1.954	3.908
4430V	2.750 ± 0.007	5.132 ± 0.007	5.258	0.40	8.89	2.687	5.375

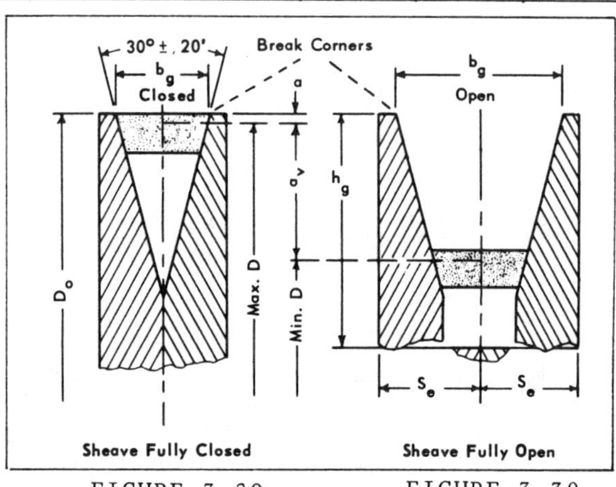

FIGURE 3.29 FIGURE 3.30

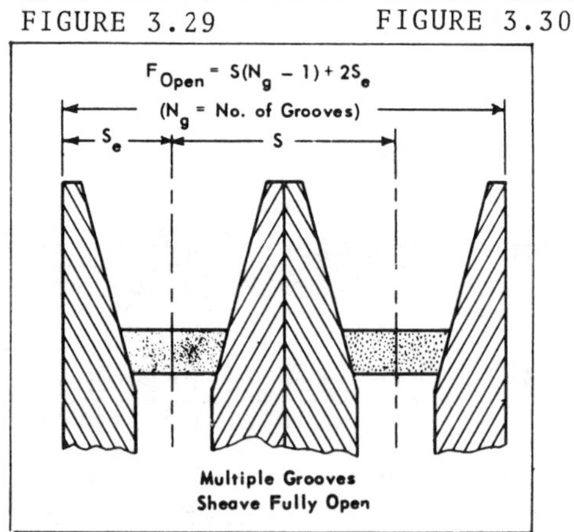

FIGURE 3.31

NOTE: D and d = Pitch diameter, large and small sheave
D_0 and d_0 = Outside Diameter, large and small sheave

Reprinted from *Engineering Standard*, Rubber Manufacturers Association, 1966, by permission of the authors.

Figure 3.32 shows a *Pulley Speed-O-Graph* for rapid calculation of the pulley laws. Using this nomograph the speed or size of either pulley can be determined when the other three factors are known.

1. Enter the chart from any given factor and follow the straight grid line to the point where it intersects— on the diagonal —the other given factor.

2. Follow the diagonal line to the point where it meets the third given factor.

3. From this point of intersection, move along the straight grid line to the fourth side of the margin for the solution.

To find the required amperage to meet a decrease or increase in rpm, use the following formula:

$$Amps_2 = Amps_1 \times \left(\frac{rpm_2}{rpm_1}\right)^3$$

Example:

A fan is turning 600 rpm at 20 amps. To deliver the proper cfm it is necessary to increase the fan speed to 700 rpm. Find the new amperage.

Solution:

Substituting in the above formula,

$$20 \times \left(\frac{700}{600}\right)^3 = 20 \times 1.58 = 31.7 \text{ amps}$$

Table 3.12 gives calculated data for this equation. Using the above example, the increase in speed is 100 rpm; this is an increase of 100/600 or 16.66%. By interpolation the table shows that the original amps would have to be multiplied by 1.58, or 20 x 1.58 = 31.7 amps.

PULLEY LAWS SPEED-0-GRAPH

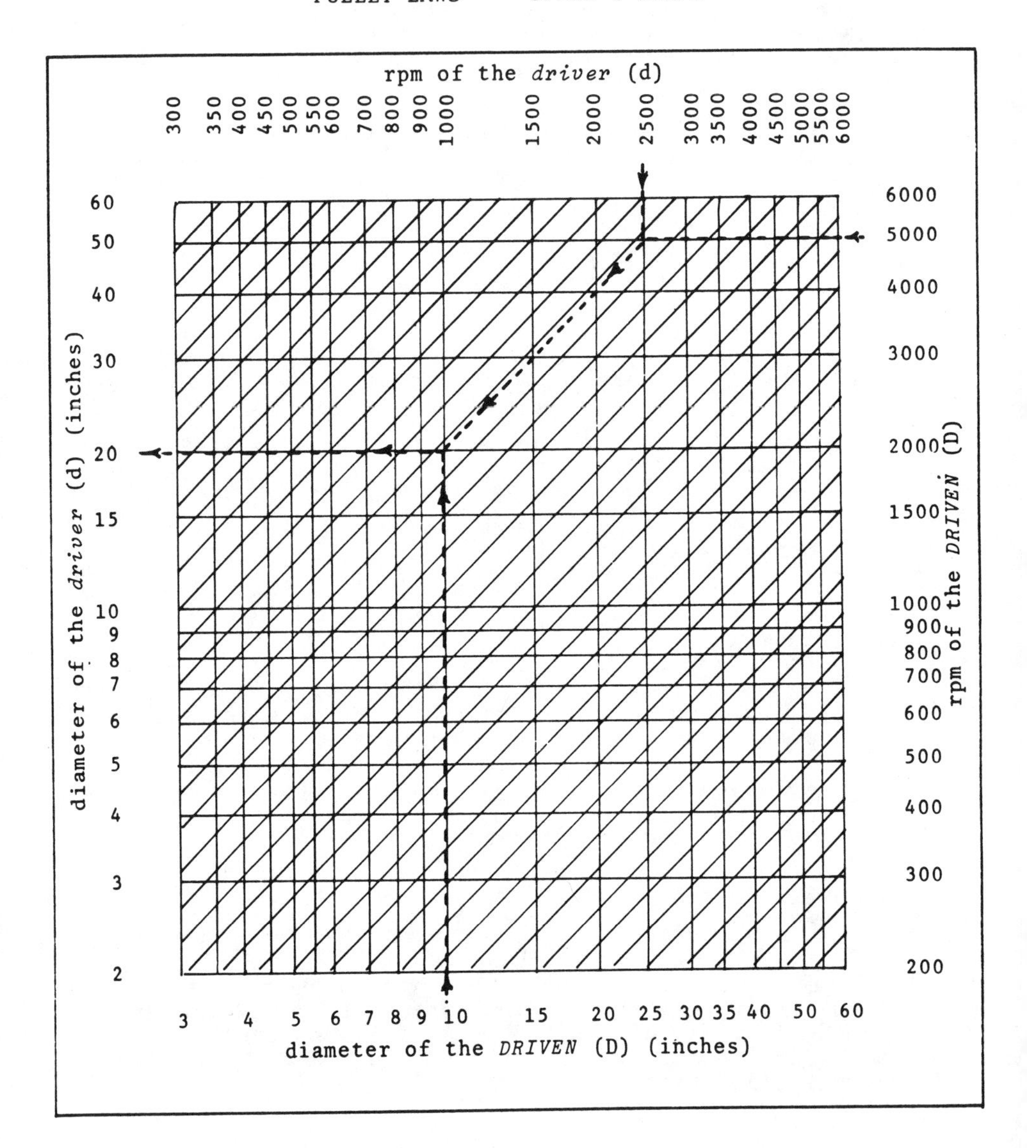

FIGURE 3.32

Example 1. Given:
Diameter of Driver = 3"
DIAMETER OF DRIVEN = 12"
rpm of Driver = 5000
FIND: RPM OF DRIVEN

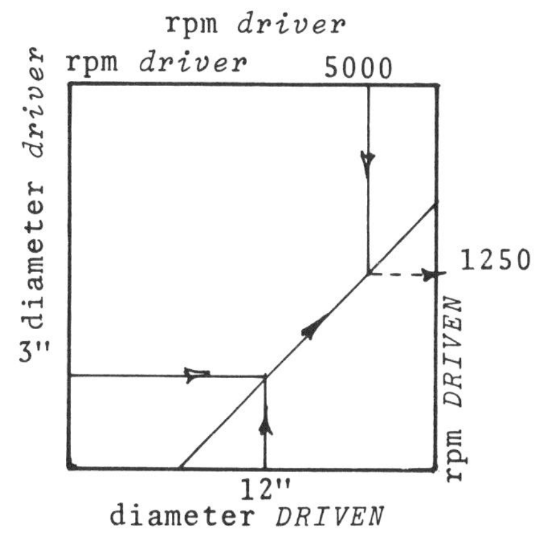

Example 2. Given:
Diameter of Driver = 30"
DIAMETER OF DRIVEN = 4"
RPM OF DRIVEN = 3750
FIND: rpm of Driver

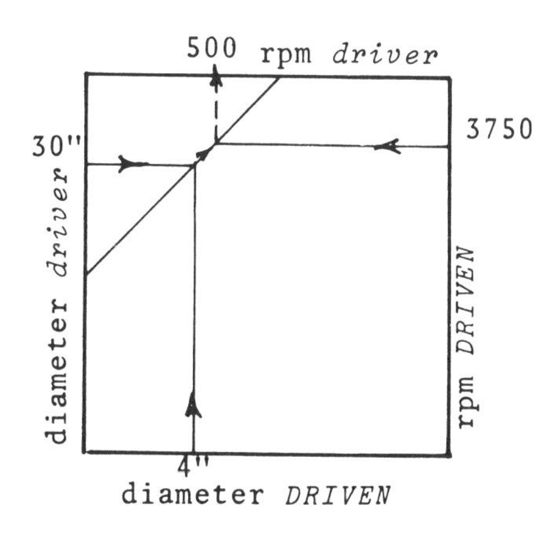

Example 3. Given:
rpm of Driver = 1000
RPM OF DRIVEN = 5000
Diameter of Driver = 10"
FIND: DIAMETER OF DRIVEN

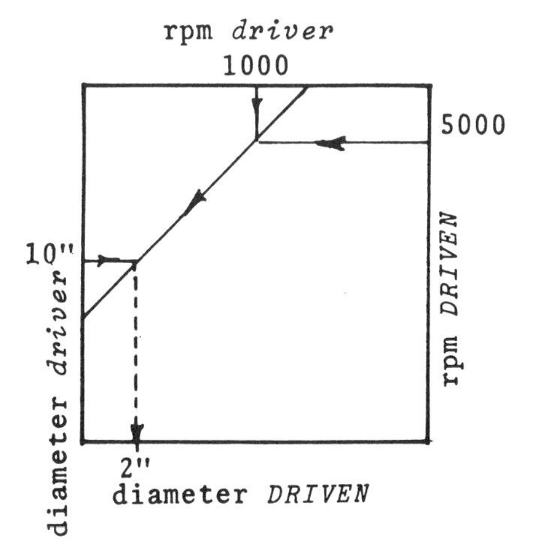

Example 4. Given:
rpm of Driver = 2500
RPM OF DRIVEN = 5000
DIAMETER OF DRIVEN = 10"
FIND: Diameter of Driver

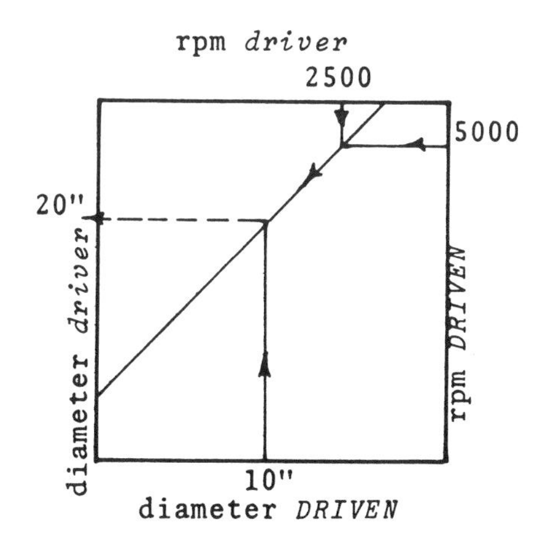

TABLE 3.12

RPM INCREASE/DECREASE

To determine the required change in fan speed multiply the measured amps by the given factor.

% Rpm Increase	Multiply Amps By:	% Rpm Decrease	Multiply Amps By:
1		1	
2	1.06	2	0.94
3	1.09	3	0.92
4	1.13	4	0.88
5	1.16	5	0.86
6	1.19	6	0.83
7	1.23	7	0.80
8	1.26	8	0.78
9	1.30	9	0.75
10	1.33	10	0.73
11	1.37	11	0.70
12	1.40	12	0.68
13	1.44	13	0.66
14	1.48	14	0.64
15	1.52	15	0.61
16	1.56	16	0.59
17	1.60	17	0.57
18	1.64	18	0.55
19	1.69	19	0.53
20	1.73	20	0.51
21	1.77	21	0.49
22	1.82	22	0.47
23	1.86	23	0.45
24	1.90	24	0.44
25	1.95	25	0.42
30	2.20	30	0.34
35	2.46	35	0.28
40	2.75	40	0.22
45	3.05	45	0.17
50	3.38	50	0.12
55	3.73	55	0.09
60	4.10	60	0.06
65	4.49	65	0.04
70	4.91	70	
75	5.36	75	
80	5.83	80	
85	6.33	85	
90	6.86	90	
95	7.41	95	

FORMULAS FOR ADJUSTING SHEAVES

1. Given a change in cfm, find the new pulley setting.

$$pd_2 = \frac{cfm_2}{cfm_1} \times pd_1$$

2. Find the maximum pulley setting to use all of the available horse power on an existing system.

$$pd_2 = \sqrt[3]{\frac{\text{maximum bhp}}{\text{bhp}}} \times pd_1$$

3. When an increase in hp is required, will the existing motor overload?

$$bhp_2 = \left(\frac{cfm_2}{cfm_1}\right)^3 \times bhp_1$$

pd = pitch diameter
bhp = brake horse power
cfm = air quantity at fan

FORMULAS FOR FINDING BHP

The brake horsepower is the horsepower *actually* required to drive a fan; it includes the energy losses in the fan but does not include the drive losses between the motor and the fan. The bhp can only be determined by actual fan test.

1. $$\text{bhp} = \frac{\text{running amps minus 1/2 no load amps}}{\text{full load amps minus 1/2 no load amps}} \times \text{nameplate hp}$$

2. $$\text{Theoretical fan bhp} = \frac{\text{cfm Tp}}{6356 \times \text{ME of fan}}$$

For most fans ME (mechanical efficiency) = 0.60

V-BELTS

All standard V-belts are identified by a standard numbering system.

Single V-belts consists of a letter-numeral combination designating length of belt. An 8.0 in. belt is designated 2L080, a 52.0 in. belt is designated 3L520, and a 100.0 in. belt is designated 4L1000. The first digit indicates the number of digits in the inch length.

Variable speed V-belts consist of a standard numbering system that indicates the nominal belt top width in sixteenths of an inch by the first two numbers. The third and fourth numbers indicate the angle of the groove in which the belt is designed to operate. The digits following the letter "V" indicates the pitch length to the nearest 1/10 in. A belt numbered 1430V450 is a V-belt of 14/16" nominal top width designed to operate in a sheave of 30° groove angle and have a pitch length of 45.0 in. Length tolerances for belts under 100 in. should be ± 0.0025 over 100 inches ± 0.0075.

The length of a V-belt is calculated by the formula:

$$L = 2C + 1.57\,(D+d) + \frac{(D-d)^2}{4C}$$

Where C = Center distance between shafts, inches
D = Outside diameter of large sheave, inches
d = Outside diameter of small sheave, inches

The center distance between two shafts is given by the formula:

$$C = \frac{K+\sqrt{K^2 - 32\,(D-d)^2}}{16}$$

Where K = 4L - 6.28 (D+d)
L = Length of belt, inches
D = Outside diameter of large sheave, inches
d = Outside diameter of small sheave, inches

Table 3.13 gives the standard dimensions of variable pitch V belts.

TABLE 3.13

STANDARD V-BELT DIMENSIONS

STANDARD BELT LENGTHS

Standard Pitch Length Designation	Standard Effective Outside Length (Inches)				
	Cross Section				
	1430V	1930V	2530V	3230V	4430V
31.5	32.1				
33.5	34.1				
35.5	36.1	36.3			
37.5	38.1	38.3			
40	40.6	40.8			
42.5	43.1	43.3			
45	45.6	45.8			
47.5	48.1	48.3			
50	50.6	50.8	50.9		
53	53.6	53.8	53.9		
56	56.6	56.8	56.9	57.1	57.3
60	60.6	60.8	60.9	61.1	61.3
63	63.6	63.8	63.9	64.1	64.3
67	67.6	67.8	67.9	68.1	68.3
71	71.6	71.8	71.9	72.1	72.3
75	75.6	75.8	75.9	76.1	76.3
80		80.8	80.9	81.1	81.3
85		85.8	85.9	86.1	86.3
90		90.8	90.9	91.1	91.3
95		95.8	95.9	96.1	96.3
100		100.8	100.9	101.1	101.3
106		106.8	106.9	107.1	107.3
112		112.8	112.9	113.1	113.3
118		118.8	118.9	119.1	119.3
125			125.9	126.1	126.3
132				133.1	133.3

NOMINAL V-BELT CROSS SECTIONS

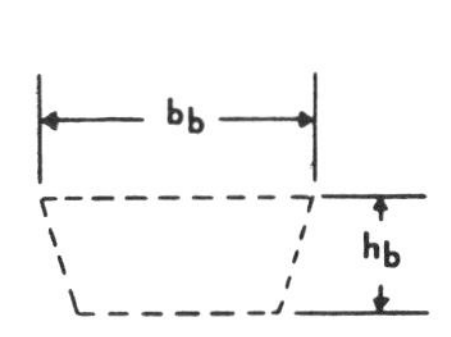

Cross Section	b_b (Inches)	h_b (Inches)
1430V	7/8	5/16
1930V	1-3/16	7/16
2530V	1-9/16	9/16
3230V	2	5/8
4430V	2-3/4	11/16

BELT SPEEDS SPEED-O-GRAPH

To find belt speed at various rpm where the sheave pitch diameter is known.

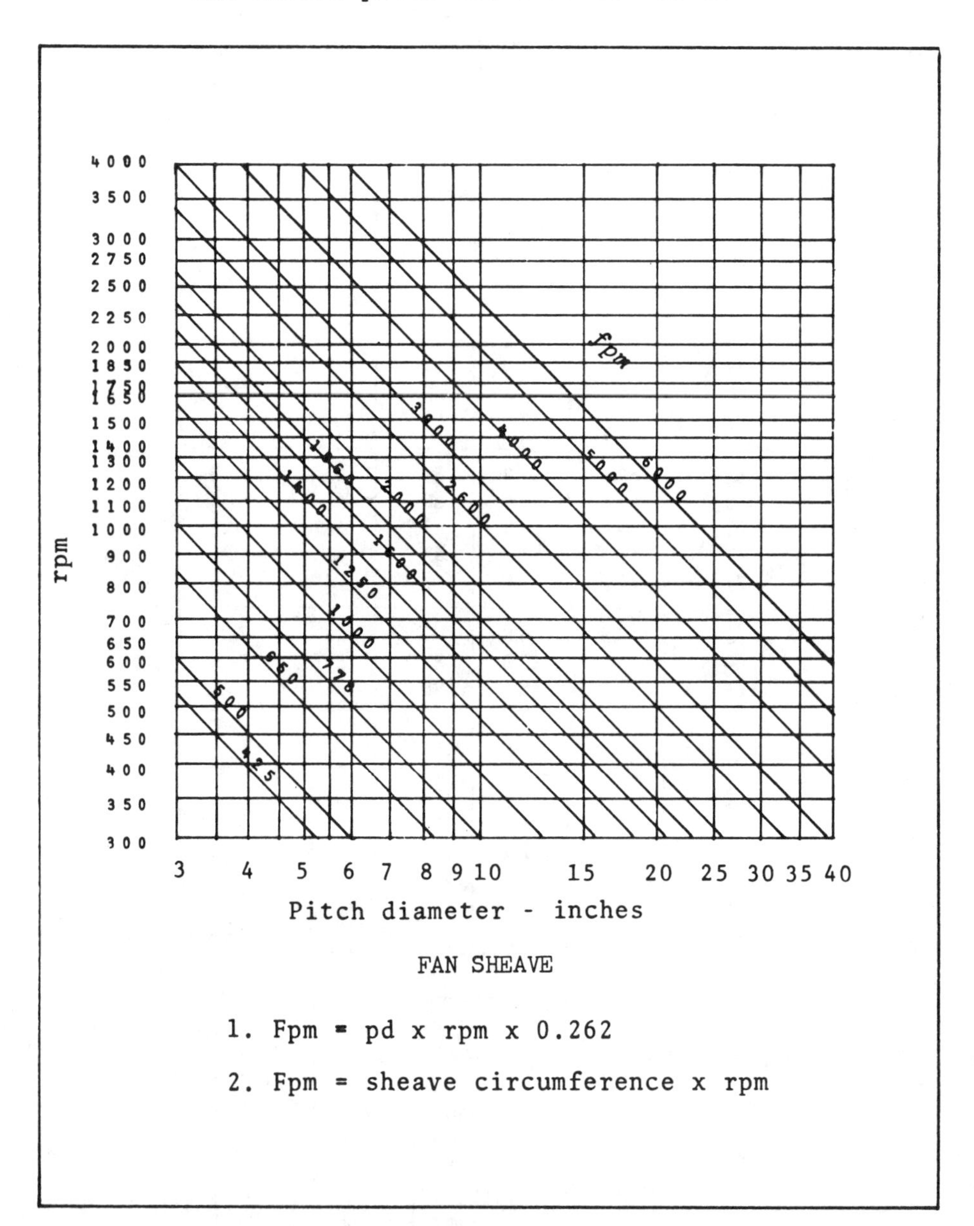

FAN SHEAVE

1. Fpm = pd x rpm x 0.262

2. Fpm = sheave circumference x rpm

FIGURE 3.33

The belt speed for various rpm's may be found where the sheave pitch diameter is known, from the formula:

fpm = pd x rpm x 0.262

Figure 3.33 gives a *Belt Speed-O-Graph* for rapid calculation of belt speeds in fpm where the pitch diameter and rpm is known. Enter at the abscissa for pitch diameter and move up to the intersecting straight grid line in rpm; the diagonal intersect will read in fpm belt speed.

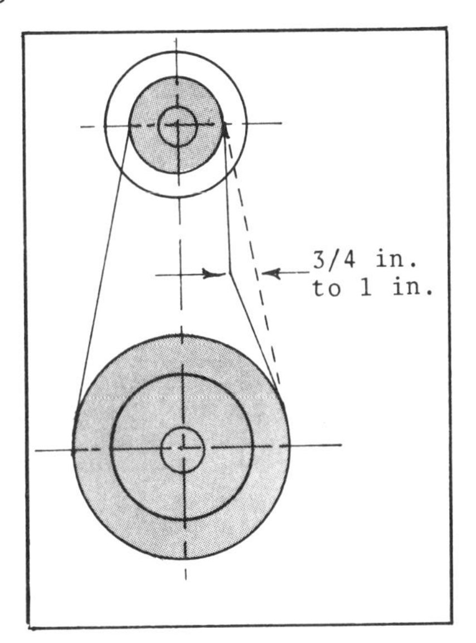

FIGURE 3.34

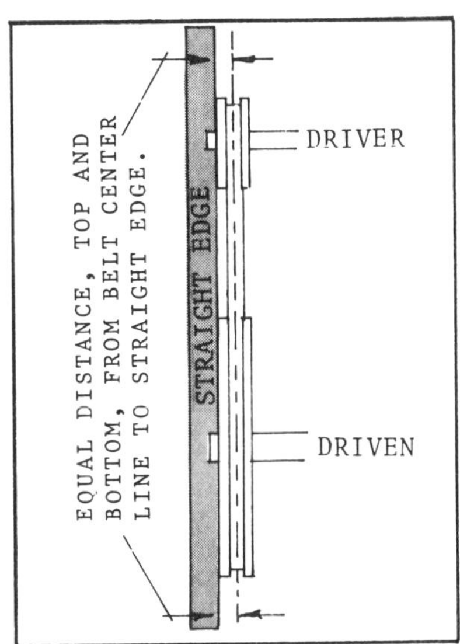

FIGURE 3.35

Belt tension adjustments should be made in three steps after the pulleys have been properly aligned.

Step 1. After initial installation of belts, run for five minutes and adjust for 3/4" to 1" deflection as shown in Figure 3.34.

Step 2. Run for 12 hours and readjust tension.

Step 3. Run for 24 hours and make final adjustment for tension.

Only matched sets of pulleys should be used and accurate alignment of the driver and driven pulley should be made as shown in Figure 3.35.

GENERAL SAFETY REQUIREMENTS

Safety regulations may vary slightly in different areas but the major basic rules are universal. The material in this section is derived mainly from three sources; *Safety Manual*, Corps of Engineers U.S. Army, *Safety Regulations*, Florida State Law 8AS, and *Manual of Crane Operation and Hitching*, Allis-Chalmers Manufacturing Company. It covers most of the material encountered in mechanical contractors' examinations as well as journeymen's.

Care and Use of Portable Ladders.-

(a) In General. To get maximum serviceability, safety, and to eliminate unnecessary damage of equipment, good safe practices in the use and care of ladder equipment must be employed by the users.

(b) Care of Ladders

(1) Ladders shall be maintained in good condition at all times, the joint between the steps and side rails shall be tight, all hardware and fittings securely attached, and the movable parts shall operate freely without binding or undue play.

(2) Metal bearings of locks, wheels, pulleys, etc., shall be frequently lubricated.

(3) Frayed or badly worn rope shall be replaced.

(4) Safety feet and other auxiliary equipment shall be kept in good condition to insure proper performance.

(5) Ladders should be stored in such a manner as to provide ease of access or inspection, and to prevent danger of accident when withdrawing a ladder for use.

(6) Wood ladders, when not in use, should be stored at a location where they will not be exposed to the elements, but where there is good ventilation. They shall not be stored near radiators, stoves, steam pipes, or other places subjected to excessive heat or dampness.

(7) Ladders stored in a horizontal position should be supported at a sufficient number of points to avoid sagging and permanent set.

(8) Ladders carried on vehicles should be adequately supported to avoid sagging and securely fastened in position to minimize chafing and the effects of road shocks. Tying the ladder to each support point will greatly reduce damage due to road shock.

(9) Wood ladders should be kept coated with a transparent protective material. Wood ladders shall not be painted with an opaque pigmental material.

(10) Ladders shall be inspected frequently and those which have de eloped defects shall be withdrawn from service for repair or destruction.

(11) Rungs shall be kept free of grease and oil.

(c) Use of Ladders

(1) Portable rung and cleat ladders shall, where possible, be used at such a pitch that the horizontal distance from the top support to the foot of the ladder is one-quarter of the working length of the ladder (the length along the ladder between the foot and the top support). The ladder shall be so placed as to prevent slipping, or it shall be lashed, or held in position. Ladders shall not be used in a horizontal position as platforms, runways, scaffolds.

(2) Crowding on ladders shall not be permitted. Portable ladders are designed as a one-man working ladder based on a 200-pound load.

(3) Portable ladders shall be so placed that the side rails have a secure footing. Safety shoes of good substantial design should be installed on all ladders. Where ladders with no safety shoes or spikes are used on hard, slick surfaces, a foot ladder board should be employed. The top of the ladder must be placed with the two rails supported, unless equipped with a single support attachment. Such an attachment should be substantial, and large enough to support the ladder under load.

(4) Ladders shall not be placed in front of doors opening toward the ladder unless the door is blocked open, locked, or guarded.

(5) Ladders shall not be placed on boxes, barrels, or other unstable bases to obtain additional height.

(6) To support the top of a ladder, at a window opening, a board should be lashed across the back of the ladder, extending across the window and providing firm support against the building walls or window frames.

(7) When ascending or descending, the user shall face the ladder.

(8) Ladders with broken or missing steps, rungs, or cleats, broken side rails, or other faulty equipment shall not be used. Improvised repairs shall not be made.

(9) Short ladders shall not be spliced together to provide long sections.

(10) Ladders made by fastening cleats across a single rail shall not be used.

(11) In building construction, where warranted by height or operations or traffic conditions, separate ladders shall be designated for ascent and descent.

(12) Improper Use. Ladders should not be used as a brace, skid, guy or gin pole, gangway, or for other uses than that for which they were intended, unless specifically recommended for use by the manufacturer.

(13) Tops of the ordinary types of step ladders shall not be used as teps

(14) On two-section extension ladders the minimum overlap for the two sections in use shall be as follows:

Size of Ladder (Feet)	Overlap (Feet)
Up to and including 36	3
Over 36 up to and including 44	4

(15) The back section of a step ladder is for the support of the front section and shall not be used for ascending or descending.

(16) Where it is necessary to install a gang ladder wide enough to permit traffic in both directions at the same time, a center rail shall be provided. One side of the ladder should be plainly marked "up" and the other side "down".

(17) Electrical hazards. All metal ladders are electrical conductors. They shall not be used in the vicinity of electrical equipment.

(18) The side rails of all portable ladders shall extend not less than three and one-half (3-1/2) feet above the platform or floor served. The ladder should be so placed that the landing rung is at or slightly above the floor or platform.

(19) Areas where portable ladders are used shall be kept clear of rubbish and waste materials. Unused materials shall be safely stored.

Boatswains' Chairs.-

(a) The chair shall be a seat not less than two feet long by one foot wide and one inch thick.

(b) Cleats shall be nailed to the underside of each end of the chair and shall project at least nine inches in front of the seat.

(c) The chair shall be supported by means of a sling attached to a suspension rope. The entire assembly shall have a safety factor of four.

(d) The suspension rope shall either be securely fastened to a fixed object overhead, or passed through an overhead block securely fastened. The free end shall be securely fastened to a fixed and easily accessible object, and the chair raised or lowered, if necessary with the aid of helpers.

(e) When the suspension rope is attached by means of a hitch, the workman shall be provided with stirrups upon which he can rest his weight while he is shifting the hitch by which the chair is made fast, and the stirrups shall be supported independently of the chair itself.

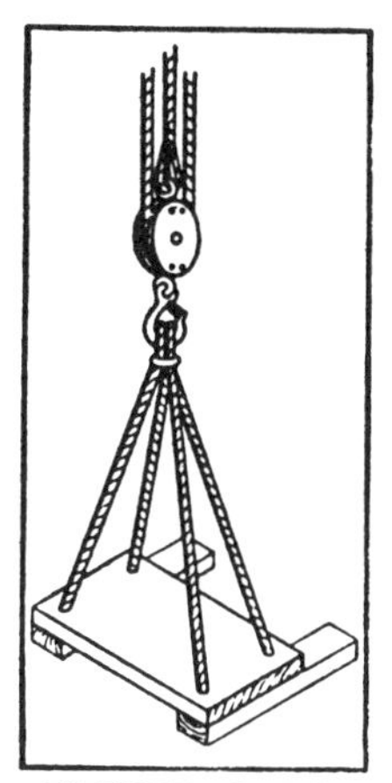

FIGURE 3.36

(f) Every workman using a boatswains' chair shall be provided with a safety belt secured to the supporting tackle.

(g) When a boatswains' chair is used by a workman using a torch, or any open flame, fiber rope slings shall not be used. The slings shall be at least three-eighths inch wire rope.

Tie Beam Jack Scaffold.-

(a) This type of scaffold shall be for light work inside buildings and shall be composed of framed wood jacks supporting a plank platform or construction of equivalent strength.

(b) The maximum height of the jacks shall be 60 inches, the base shall be not less than 30 inches and the top member not less than 20 inches long.

(c) If the jacks are framed up of lumber, the top member, the inside upright (the one against the wall) and the outside sloped member shall be of 2 x 4 inch lumber and the bottom member of 1 x 6 inch boards.

(d) At the top of the jack, in addition to the 2 x 4 member, two 1 x 6 inch boards shall be nailed parallel to the top member and on its sides as bearers.

(e) A 1 x 6 inch diagonal brace shall extend from near the bottom of the upright member to near the top of the outside sloped member.

(f) Platform planks shall be nailed to the top member of each jack.

(g) A 1 x 4 inch logitudinal brace shall be nailed along the outside of each jack at floor level.

(h) Jacks shall be secured to the walls of the building to prevent toppling and wedged against end walls to prevent lateral displacement.

(i) The jacks shall be set up not more than eight feet apart.

(j) When this type of scaffold is used, it shall not be built up more than one jack high and the scaffold shall be set up on a level and stable foundation.

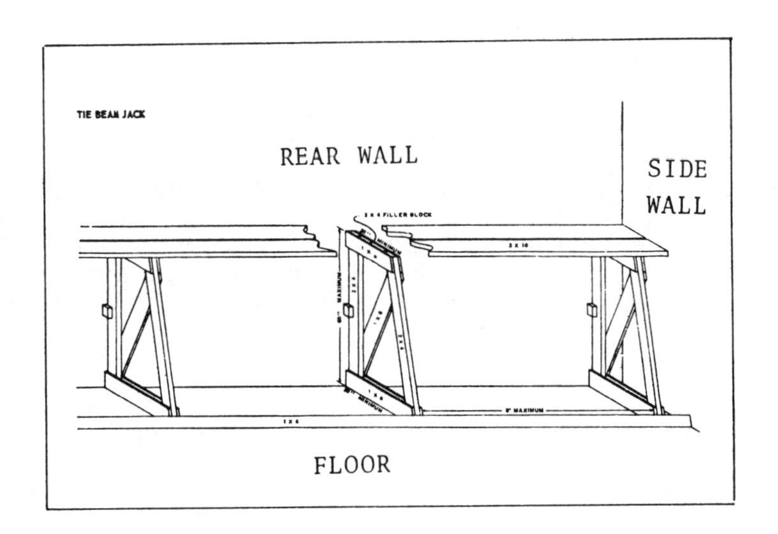

FIGURE 3.37

Hand Signal for Crane Operation

Figure 3.38 gives the standard hand signals for crane operation as recommended by The American Society of Mechanical Engineers, *Standard B30.5-1968, Crawler, Locomotive and Truck Cranes.* Reprinted by permission of the publisher, The American Society of Mechanical Engineers, 345 East 47th Street, New York.

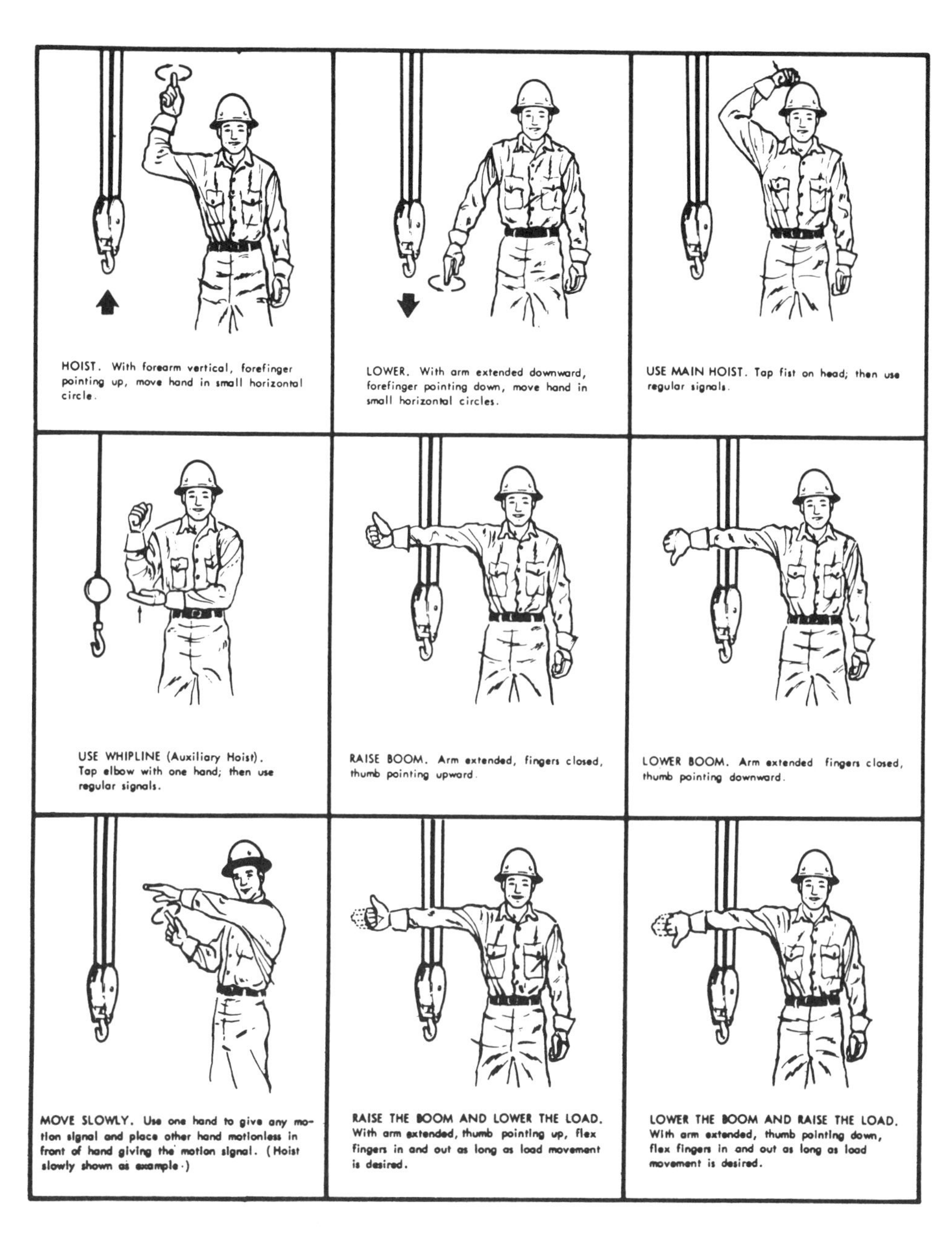

FIG. 1 STANDARD HAND SIGNALS FOR CONTROLLING CRANE OPERATIONS

FIGURE 3.38

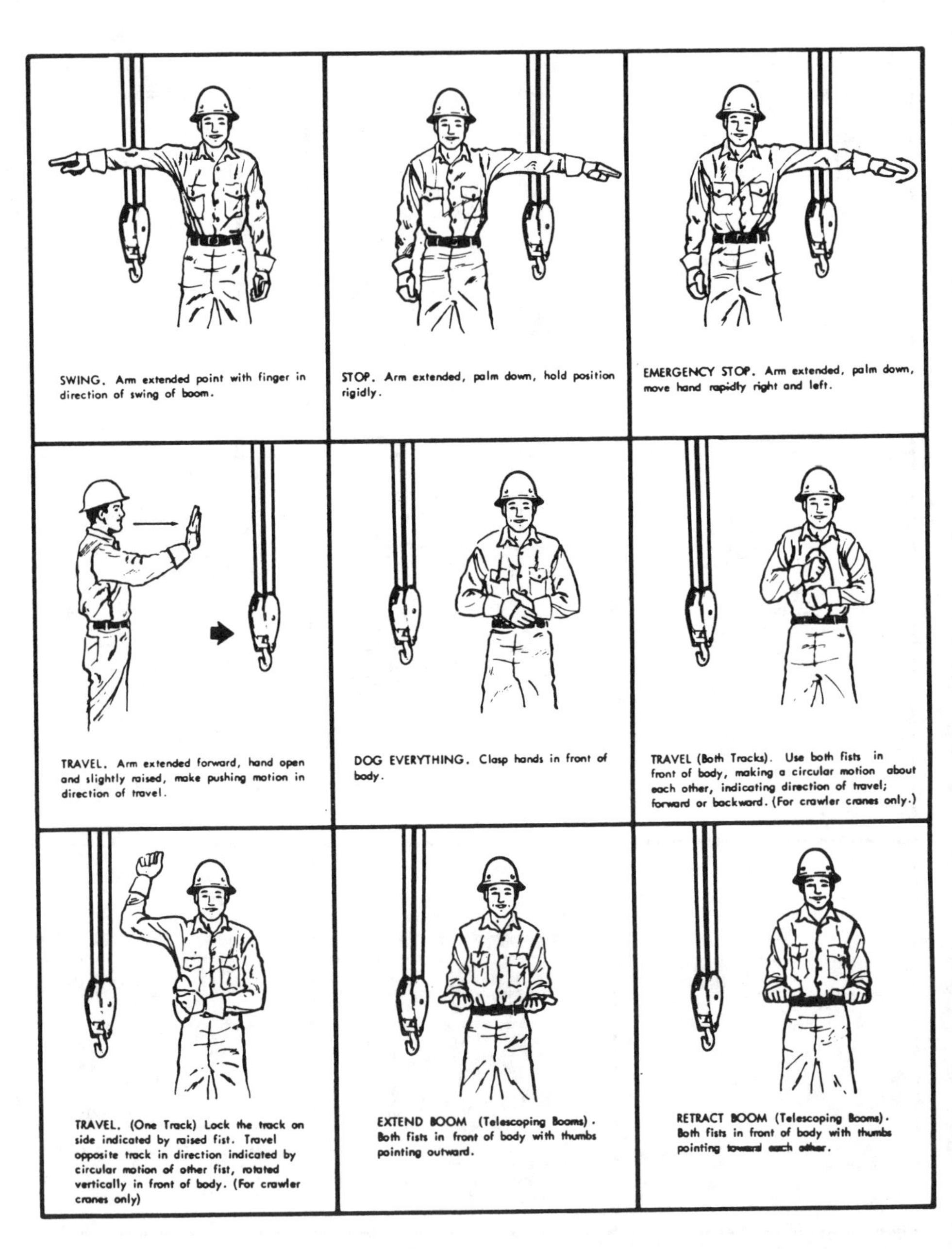

FIG. 1 STANDARD HAND SIGNALS FOR CONTROLLING CRANE OPERATIONS

FIGURE 3.38

Hitching Equipment and Safe Practice

When using chokers do not pass the running line through an eye splice - use a shackle, and keep the shackle pin in the eye.

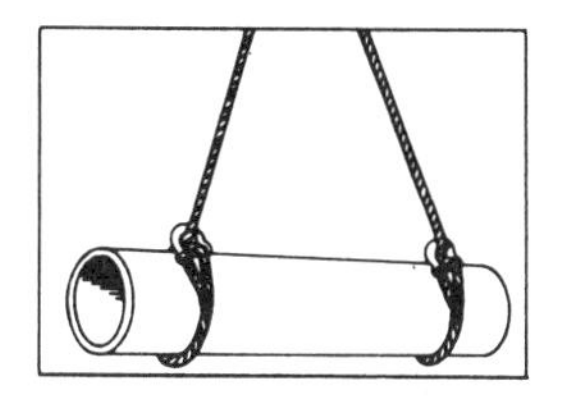

Use shackles for wire rope choker hitches.

FIGURE 3.39

When forming an eye with clamps always have the nuts on the standing part and use a thimble as shown in Figure 3.40.

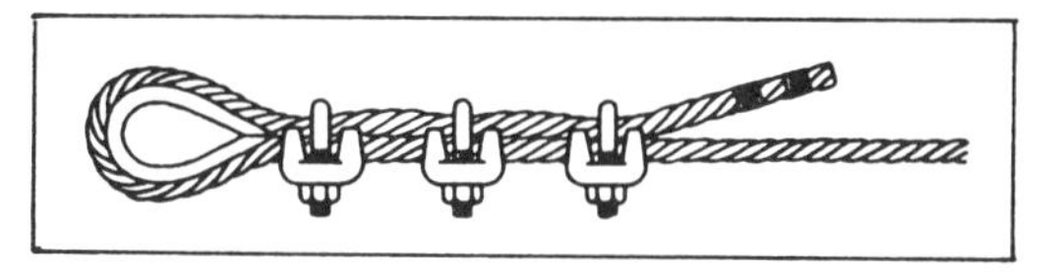

Use of thimble in eye splice.

FIGURE 3.40

Wire rope slings should be protected from sharp corners with corner irons or heavy chafing gear, Figure 3.42.

Each leg of a wire rope sling should be secured at the hook to prevent reeving of the sling on the hook. The proper method of rigging wire rope slings to the hook is shown in Figure 3.41.

Wire rope should never be knotted and the wire should be visually checked each time before using.

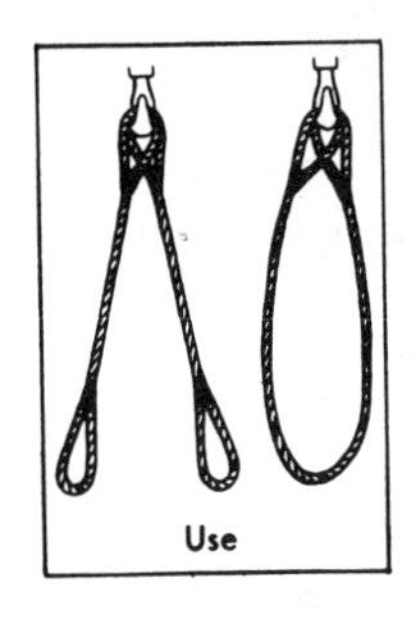

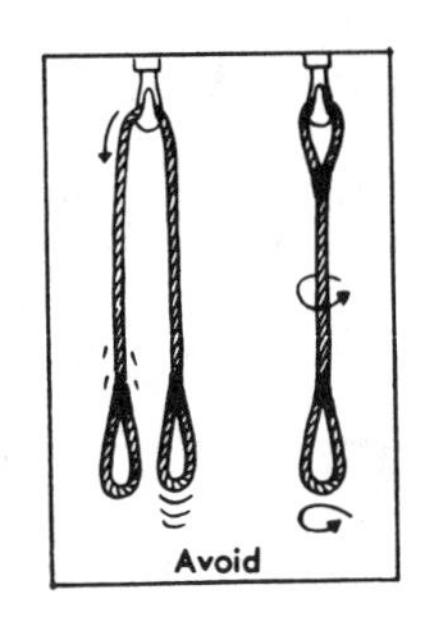

FIGURE 3.41

FIGURE 3.42

Manila rope should be protected against weather, excessive heat, solvents, grease etc., and should be visually inspected each time before using. Manila slings should be protected with chafing gear or bagging as shown in Figure 3.42. A blackwall hitch is shown in Figure 3.43 this is the proper method of rigging a manila rope sling to the hook. The proper steps in making a wrap-around hitch are illustrated in Figure 3.44.

FIGURE 3.43

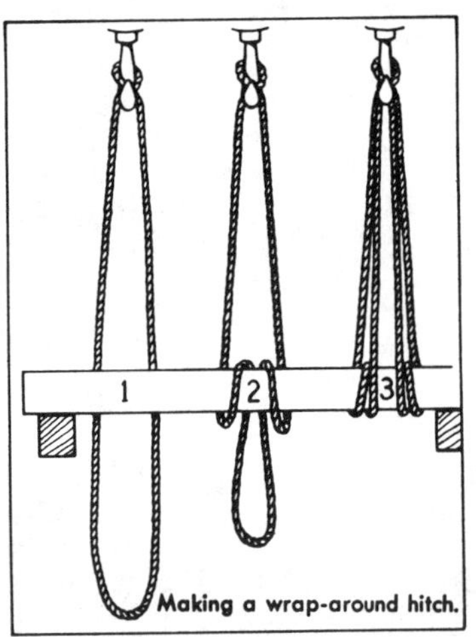

FIGURE 3.44

Safe Load Tables

Safe loading for manila and wire rope slings is a function of the number of sling legs connecting the crane hook, and the load angle; the smaller the load angle, the lower the lifting efficiency. Table 3.14 gives the angle between the horizontal surface of the load and the sling as a percentage of a straight lift.

TABLE 3.14

SLING ANGLE EFFICIENCY

LOAD ANGLE Degrees	LIFTING EFFICIENCY Percent
90	100
60	86.6
45	70.7
30	50
15	25
0	0

To use the following Safe Load Tables, see Figure 3.45. Assuming a packaged chiller weighs 32,000 lbs and the load angle will be 30°; from Table 3.17 use a 1-1/2 in. diameter wire rope sling with four legs. Now, dividing the load by four shackles gives four tons per shackle, but Table 3.44 shows that the lifting efficiency for a 30° angle is 50%. Therefore, *double* the listed 90° safe load in Table 3.18 and select 1-1/4 in. shackles.

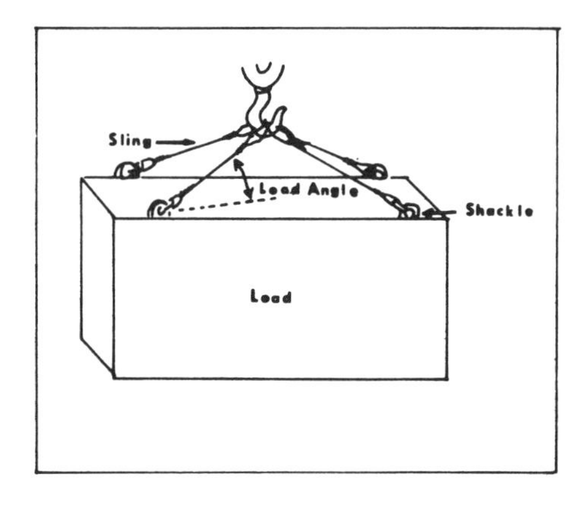

TABLE 3.45

TABLE 3.15

MANILA ROPE SLINGS - STRAIGHT LEG

ROPE DIAMETER INCHES	Straight 2 Leg	Straight 4 Leg	Choker 2 Leg	60° Choker 4 Leg	45° Choker 4 Leg	30° Choker 4 Leg
	TONS					
1/2	1/2	1	1/3	2/3	1/2	1/3
3/4	3/4	1-1/2	3/4	1-1/4	1	3/4
1	1-1/2	3	1-1/4	2	1-1/2	1-1/4
1-1/2	3	6	2	4	3	2
2	5	10	4	7	6	4
2-1/2	7	15	6	10	8	6
3	10	20	8	14	12	8
3-1/2	14	29	11	20	16	11
4	17	34	13	23	19	13

TABLE 3.16

MANILA ROPE SLINGS — BASKET

ROPE DIAMETER INCHES	60° Basket 4 Leg	45° Basket 4 Leg	30° Basket 4 Leg	60° Basket 6 Leg	45° Basket 6 Leg	30° Basket 6 Leg
	TONS					
1/2	3/4	2/3	1/2	1	3/4	2/3
3/4	1-1/2	1	3/4	2-1/4	2	1
1	2-1/2	2	1-1/2	4	3	2
1-1/2	5	4	3	8	6	4
2	9	7	5	13	11	7
2-1/2	13	11	7	19	16	11
3	18	15	10	27	22	15
3-1/2	25	21	15	38	31	22
4	30	24	17	44	36	25

TABLE 3.16

WIRE ROPE SLINGS — STRAIGHT LEG

ROPE DIAMETER INCHES	Straight 1 Leg	Choker 1 Leg	60° Choker 2 Leg	45° Choker 2 Leg	30° Choker 2 Leg
	TONS				
1/4	1/2	1/3	2/3	1/2	1/3
3/8	1	3/4	1-1/4	1	3/4
1/2	2	1-1/2	2-1/2	2	1-1/2
5/8	3	2	4	3	2
3/4	4	3	5	4	3
1	7	5	8	7	5
1-1/4	10	7	12	9	7
1-1/2	13	9	16	13	9
2	21	15	27	22	15
2-1/2	28	22	38	31	22
3	36	28	49	40	28
3-1/2	40	34	59	48	34

TABLE 3.17

WIRE ROPE SLINGS——BASKET

ROPE DIAMETER INCHES	60° Basket 2 Leg	45° Basket 2 Leg	30° Basket 2 Leg	60° Basket 4 Leg	45° Basket 4 Leg	30° Basket 4 Leg
	TONS					
1/4	2/3	1/2	1/3	1	1	3/4
3/8	1-1/2	1	3/4	3	2	1-1/2
1/2	2-1/2	2	1-1/2	5	4	3
5/8	4	3	2	7	6	4
3/4	5	4	3	11	9	6
1	9	7	5	18	15	10
1-1/4	13	11	7	26	21	15
1-1/2	17	14	10	35	28	20
2	27	22	15	53	44	31
2-1/2	38	31	22	75	61	43
3	49	40	29	97	80	56
3-1/2	59	49	34	118	97	68

TABLE 3.18
SAFELOAD FOR SCHACKLES

Size dia.

Size	Safe Load at 90°
Inches	Tons
¼	⅓
5/16	½
⅜	¾
7/16	1
½	1½
⅝	2
¾	3
⅞	4
1	5½
1⅛	6½
1¼	8
1⅜	10
1½	12
1¾	16
2	21
2¼	27
2½	34
2¾	40
3	50

*Note: When attachments (Shackles, "S" Hooks, etc.) are used at angles other than 90°, reduce safe load rating according to Angle Efficiency Table,

Knots

The origin of most knots, hitches and bends, are lost in the dim antiquity of our seafaring ancestors. Aside from exhibits of the art of seamanship and Boy Scout jamborees, most knots and fancy rope work are now merely of academic interest. Most manuals illustrate dozens of different knots and ties, while in fact, the pipefitter — or rigger — using modern equipment, will seldom need to use more than four knots.

The *bowline* is the king of knots If made correctly it can *never* jamb. Useful for lowering buckets, tools or men over the side or whenever an absolutely safe hitch is needed. Figure 3.46.

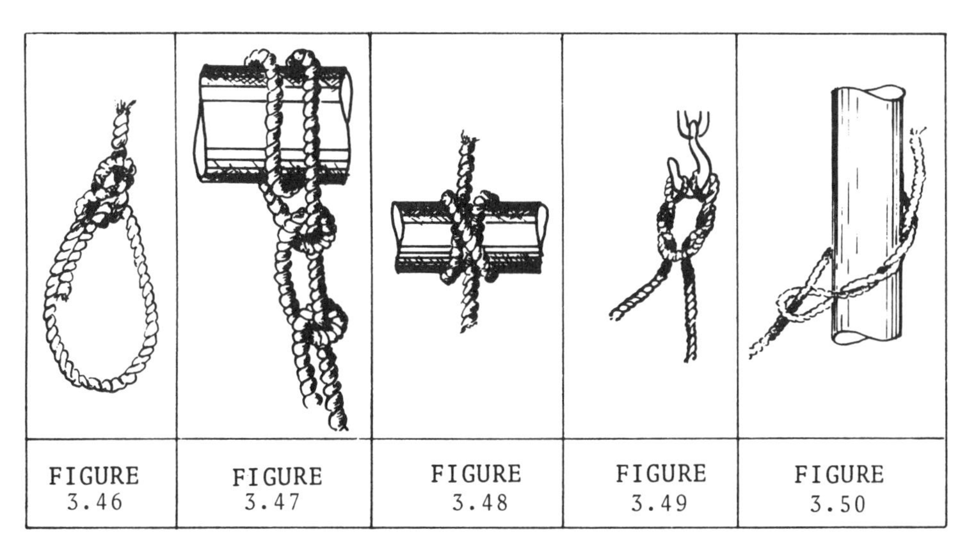

FIGURE 3.46 | FIGURE 3.47 | FIGURE 3.48 | FIGURE 3.49 | FIGURE 3.50

The *half hitch, two half hitches,* and the *round turn with two half hitches,* are very useful knots to keep the ends of pipe or timber from spinning, to hold upright objects in line with the pull, and to make fast lines of moderate size. Figure 3.47.

The *clove hitch* is really two half hitches around a stancion or pipe. A very useful knot, it will take great strain and is easy to unbend. Figure 3.48.

The *cat's paw* can be made anywhere on the bight of a rope when it is necessary to clap on a handy billy, or tackle, to the hauling part of a larger purchase. Figure 3.49.

The *timber hitch,* used together with a *half hitch* is good for towing a spar, and hoisting pipe or timber. Figure 3.50.

Grounding of Portable Electrical Equipment

All portable electric tools are to be grounded prior to use. The preferable grounding device shall be so arranged that the ground connection will be established automatically when the portable cable is connected to the fixed wiring. The grounding conductor shall be contained in or on the cable or cord assembly. The grounding conductor shall be contained in or on the cable or cord assembly. The grounding conductor shall be connected to ground, but shall not be a grounded circuit conductor.

All exposed non-current-carrying metal parts of portable electrical equipment operated at more than 50 volts to ground shall be effectively grounded regardless of use or location.

Exception - Electric tools provided with double insulation are exempted.

Table 5.7 offers a guide to the size of conductors for portable tools.

TABLE 3.19

CONDUCTORS FOR PORTABLE POWER TOOLS

HP	50'	100'	150'	200'	300'
1 or less	16	16	14	12	12
1-1/2	14	14	12	10	8
2	10	10	8	8	6
3	8	8	6	6	4

Eye Protection

Eye protection is required whenever there is reasonable probability of injury, and the use of suitable eye and/or face protectors is mandatory when exposed to any of the following:

Arc welding, heavy gas cutting, scarfing.
Gas welding, brazing and cutting.
Machinery or grinding of any materials causing flying particles.
Spot welding.
Powder actuated tools.
Abrasive blasting.
Injurious radiations.

PART 4

QUIZZES & TEST

BASIC REFRIGERATION QUIZ NO.1

1. The definition of refrigeration is

   A. Automatic temperature control.
   B. Removal of heat.
   C. Absolute enthalpy.
   D. Adibatic rejection.
   E. None of the above.

2. The definition of cold is

   A. A condition of low temperature
   B. Removal of heat.
   C. 12,000 Btu per hour.
   D. Low dew point.
   E. High humidity.

3. The definition of sensible heat is

   A. Changing the state of a substance.
   B. 144 Btu per pound.
   C. Changing the temperature of a substance.
   D. Heat flow from one substance to another.

4. The definition of latent heat is

   A. Changing the state of substance
   B. 144 Btu per pound.
   C. Changing the temperature of a substance
   D. Heat flow from one substance to another.

5. The definition of a Btu is

   A. The amount of heat to raise one pound of water one degree F.
   B. The anount of heat to raise one pound of water one degree C.
   C. The amount of heat rejection at the condenser.
   D. Enthalpy rise and fall.

6. One ton of refrigeration in terms of Btu removal per hour is equal to

   A. 144 B. 970 C. 4350 D. 12,000 E. 25, 500

7. The amount of heat required to change one pound of water to one pound of steam at atmospheric pressure is

   A. 144 Btu B. 970 Btu C. 4350 Btu D. 12,000 Btu E. 24,500 Btu

8. Dichlorodifluoromethene CCL_2F_2 is

   A. R 11 B. R 12 C. R 13B1 D. R 115 E. R 500

9. Ammonia is a refrigerant of group

   A. 1 B. 2 C. 3 D. 4 E. 5

10. Propylene as a refrigerant is considered

    A. Toxic B. Flammable C. Florocarbon D. Gastric E. Absorptive

BASIC REFRIGERATION QUIZ NO. 2

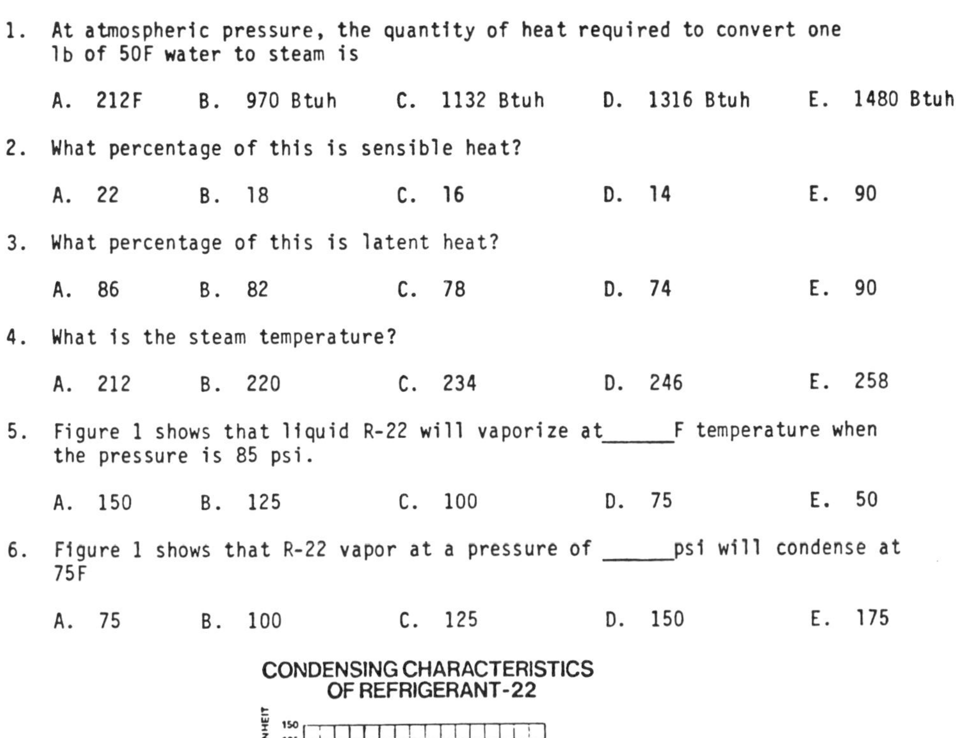

1. At atmospheric pressure, the quantity of heat required to convert one lb of 50F water to steam is

   A. 212F B. 970 Btuh C. 1132 Btuh D. 1316 Btuh E. 1480 Btuh

2. What percentage of this is sensible heat?

   A. 22 B. 18 C. 16 D. 14 E. 90

3. What percentage of this is latent heat?

   A. 86 B. 82 C. 78 D. 74 E. 90

4. What is the steam temperature?

   A. 212 B. 220 C. 234 D. 246 E. 258

5. Figure 1 shows that liquid R-22 will vaporize at______F temperature when the pressure is 85 psi.

   A. 150 B. 125 C. 100 D. 75 E. 50

6. Figure 1 shows that R-22 vapor at a pressure of ______psi will condense at 75F

   A. 75 B. 100 C. 125 D. 150 E. 175

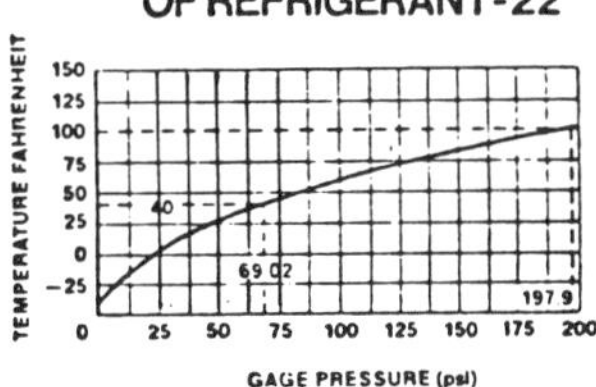

FIGURE 1

7. What determines whether a liquid refrigerant will vaporize or a refrigerant vapor will condense at a given pressure?

   A. A change in the dew point characterstic.
   B. The addition of 212F or the removal of 32F.
   C. The addition of heat adequate to reach the condensing characteristics.
   D. The removal of heat necessary to reach the vaporization characteristics.
   E. The addition or removal of heat.

8. According to the Trane AC Manual 10 lb of R-12 under a pressure 206 psig will boil at approximately__________F.

   A. 134 B. 136 C. 138 D. 140 E. 142

9. The total heat content of the vapor in Question 8 is_________Btuh (enthalpy).

   A. 89.588 B. 895.88 C. 89.967 D. 899.67 E. 90.204

10. In a standard mechanical refrigeration system the refrigerant gas pressure is usually raised to _________times higher in the pumping process.

    A. 1 to 2 B. 2 to 4 C. 4 to 6 D. less than one

    E. None of the above

COOLING TOWER QUIZ NO. 3

Given: A 30 ton atmospheric spray type cooling tower on a roof. Vertical distance from the pump to the tower spray header = 50 ft, vertical distance from tower sump return to the pump = 40 ft. System layout in Figure below.

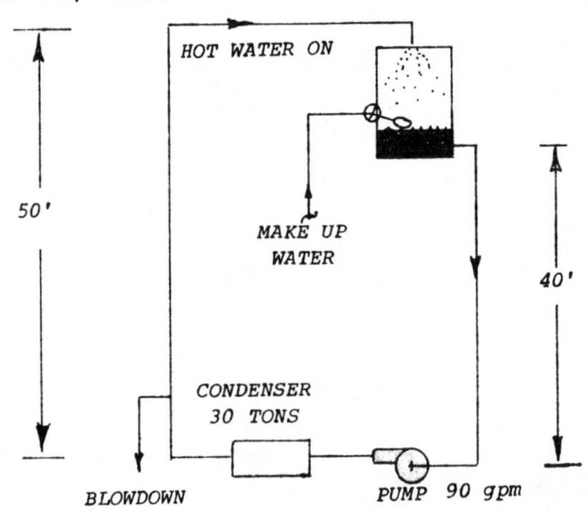

Find: The total static head.

A. 23.1 psi
B. 90 ft
C. 50 ft
D. 4.33 psi

COOLING TOWER QUIZ NO. 4

Given: A water tank on an office building has its surface 260 ft above ground. All faucets are closed (no water drain off). There is a pressure gauge at the ground level.

Find: The gauge reading at the ground.

A. 112.58 psi
B. 164.22 psi
C. 197.33 psi
D. 225.16 psi

COOLING TOWER QUIZ NO. 5

Given: The system shown in Figure below.

Find: The actual head in feet of water.

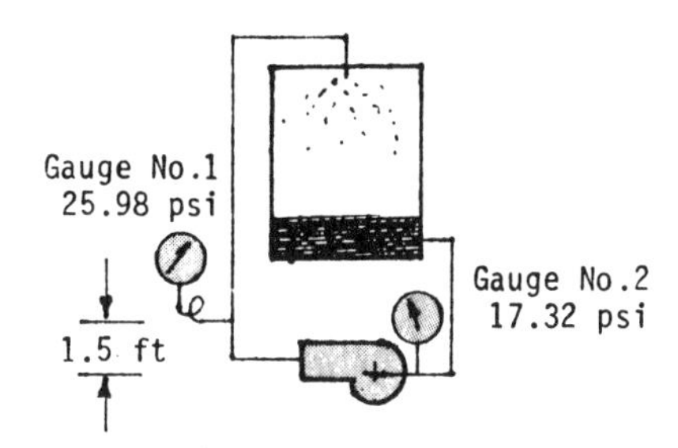

MATH QUIZ NO. 6

(One-half hour calculator quiz)

ADD:

1. $251.05 + 33.582 + 6953.01 + 52 + .03 =$
2. $953.115 + 66.05 - 57.32 + 63.1 - 362.132 =$
3. $-2 + (-8) + (-6) =$
4. $-8 + 6 + (-7) + 4 + 16 + (-2) =$

MULTIPLY:

5. $90 \times 3 \times 5.27 =$
6. $(-90) \times (-3) =$
7. $(-90) \times (+3) =$
8. $.245 \times .78 \times 1.35 =$
9. $(-30) \times (-8) \times (+4) \times (-20) =$

DIVIDE:

10. $\frac{120}{4} =$
11. $\frac{-426}{4} =$
12. $\frac{520 \times 123}{43560} =$
13. $\frac{.333 \times 49 \times 87}{27} =$
14. $\frac{\pi \times 90 \times 50}{37 \times 10} =$
15. $\frac{(8 + 27) \times 60 \times .0175}{6 \times 27 \times .10} =$

CHANGE THE FOLLOWING TO FEET;

16. 2' 4" =______ 25' 3" =______ 11" =______ 5'1" =______

MULTIPLY AND GIVE ANSWER IN FEET;

17. 6' 9" x 112' 2" =
18. 16' 3" x 8' 6" x 3' 4" =

COMPRESSOR TROUBLE SHOOTING QUIZ NO. 7

In each of the following questions, one answer is best. Which answer is the best?

1. Evaporator freezes but defrosts while the compressor is running.

   A. Condenser water too cold
   B. Thermostat setting too low
   C. Superheat adjusted too low
   D. System is undercharged
   E. Moisture in system

2. Space temperature will not pull down.

   A. Wiring incorrect
   B. Control contact stuck
   C. Loose thermostat bulb
   D. Expansion valve clogged or defective
   E. Superheat adjusted too low

3. Running capacitor burns out.

   A. Low voltage
   B. High voltage
   C. Wrong type relay
   D. Wrong start capacitor
   E. Control set too high

4. Compressor short cycles.

   A. Moisture in system
   B. Clogged expansion valve
   C. Restricted discharge line
   D. Leaking discharge valve
   E. Start winding held in circuit too long

5. Head pressure too low.

   A. Superheat adjusted too low
   B. System is undercharged
   C. Control differential set too close
   D. Expansion valve too small
   E. Expansion valve open too wide

6. Head pressure too high.

   A. Evaporator oil clogged
   B. Evaporator coil too small
   C. Solenoid valve leaking
   D. High voltage
   E. System is overcharged

7. Starting capacitor burns out.

   A. High voltage
   B. Low voltage
   C. Control set too high
   D. High pressure control cutting out
   E. Overload protector cutting out

8. Starting relay burns out.

   A. solenoid valve leaking
   B. Thermostat bulb loose
   C. Discharge valve leaking
   D. High pressure control cutting out
   E. Low voltage

9. Suction line frosting.

   A. Expansion valve too small
   B. Solenoid valve leaking
   C. Condenser water too cold
   D. Expansion valve clogged or restricted
   E. Superheat adjusted too low

COMPRESSOR TROUBLE SHOOTING QUIZ NO. 8

In each of the following questions, one answer in the answer band is incorrect. Which answer is incorrect?

1. Compressor will not start; does not hum.

   A. Defective relay
   B. Defective compressor motor
   C. Blown fuse
   D. Defective overload protector
   E. Overload protector tripped

2. Compressor will not start. It hums but cycles on overload protector.

   A. Low voltage
   B. Grounded compressor
   C. Defective run capacitor
   D. Weak start capacitor
   E. Relay contacts stuck

3. Compressor starts and runs but cycles on overload.

   A. Low voltage
   B. Grounded compressor
   C. Fan or pump motor wired to the wrong side of the overload protector
   D. Defective compressor motor
   E. Unbalanced 3ϕ voltage

4. Compressor starts but cuts right out on overload.

   A. Defective overload protector
   B. Low voltage
   C. Relay contacts open
   D. Defective start capacitor
   E. Defective run capacitor

5. Liquid line is hot.

   A. Restricted capilary tube
   B. Overcharged system
   C. Expansion valve open too wide
   D. Air in system
   E. High head pressure

6. Unit is noisy.

   A. Compressor oil charge too low
   B. Refrigerant charge too low
   C. Fan blade out of pitch
   D. Fan motor bearings worn
   E. Condensing unit part loose

7. Compressor starts, but starting winding stays in circuit.

   A. Defective overload protector
   B. Wrong run capacitor
   C. Wrong start capacitor
   D. Compressor grounded
   E. High head pressure

8. When thermostat closes compressor starts but repeatedly cuts out on overload. After several such attemps, the compressor finally starts.

   A. Restriction at the drier
   B. Restriction at the capillary tube
   C. Thermostat differential set too close
   D. Relay contacts badly pitted
   E. Unbalanced 3ϕ voltage

9. Unit does not cut off or runs too long.

   A. Wrong type relay
   B. System undercharged.
   C. Expansion valve leaking
   D. Equipment too small for the load
   E. Defective or clogged expansion va

SERVICE AND MAINTENANCE QUIZ NO. 9 Section 1

1. What will the readings of wet and dry hygrometers be if exposed to air that is completely saturated:

    (a) They will both read alike

    (b) The wet bulb will be lower

    (c) The dry bulb will be lower

    (d) 90% relative humidity

2. Type K copper tube has a greater wall thickness than Type L copper tube:

    (a) True (b) False

3. What type of electric motor has a set of field coils and a rotating armature:

    (a) An induction motor

    (b) A capacitor split phase motor

    (c) A repulsion start indiction run motor

    (d) Dual winding repulsion induction motor

4. When air is passed through a water spray some of the moisture of the air is given up only if the temperature of the water is:

    (a) Below the dewpoint of the air

    (b) Below the temperature of the air

    (c) Above the dewpoint of the air

    (d) Above the temperature of the air

5. On a magnetic starter the alloy piece holding an overload relay closed which melts when current drawn is too great, is known as the thermal overload--

    (a) Holding coil (c) Relay

    (b) Heater (d) Element

6. What factors determine the size of an expansion tank:

    (a) The amount of space the water in the system requires in its expanded state

    (b) The amount of space the air in the system requires in its expanded state

    (c) The amount of air in system

    (d) The operating pressure in system

7. What is the gross weight of refrigerant cylinder:

    (a) The weight of the refrigerant plus the weight of the cylinder

    (b) The weight of the cylinder

    (c) The weight of the contents only

    (d) The weight of cylinder and line

8. What does an anemometer measure?

    (a) Feet (c) cu ft

    (b) Feet per minute (d) cu ft per minute

9. The term "induced draft" could refer to:

    (a) A type of control

    (b) A type of compressor

    (c) A type of cooling tower

    (d) A type of diagram

10. What fitting should be used to connect 1/4" O.D. copper tubing to a 1/4" internal pipe thread opening in the compressor body?

    (a) A union (c) A tee fitting

    (b) A half-union (d) A street ell

SERVICE AND MAINTENANCE QUIZ NO. 9 Section 2

11. The ratio of the weight of water vapor associated with a pound of dry air to the weight of water vapor associated with a pound of dry air saturated at the same temperature:

    (a) Specific humidity (c) Absolute humidity
    (b) Relative humidity (d) Percentage humidity

12. An instrument for measuring pressure, essentially a U-tube partially filled with a liquid, usually water mercury, or a light oil, so constructed that the amount of displacement of the liquid indicates the pressure being exerted on the instrument:

    (a) Potentiometer (c) Manometer
    (b) Volometer (d) Anemometer

13. A piping system in which the heating or cooling medium from several heat transfer units is returned along paths arranged so that all circuits composing the system or composing a major subdivision of it are of practically equal length.

    (a) Reversed return (c) Two pipe
    (b) Indirect (d) Down feed

14. On a steam system what valve should be used in a horizontal line where drainage is required?

    (a) Any valve (c) Gate valve
    (b) Globe valve (d) Check valve

15. How much should the overall drop (pitch) be for a 50 foot steam main?

    (a) 2-1/2" (c) 1"
    (b) 3" (d) 5"

16. In the discussion of the refrigeration cycle, Compression Ratio is expressed in the following equation thus:

    (a) Compression ratio - $\frac{\text{Discharge pressure psi absolute}}{\text{Suction pressure psi absolute}}$

    (b) Compression ratio - $\frac{\text{Suction pressure psi absolute}}{\text{Discharge pressure psi absolute}}$

    (c) Compression ratio - $\frac{\text{Discharge temperature absolute}}{\text{Suction temperature absolute}}$

    (d) Compression ratio - $\frac{\text{Suction temperature absolute}}{\text{Discharge temperature absolute}}$

17. Pressure limiting devices shall be provided on all systems operating above atomospheric pressure and containing more than how many pounds of refrigerant:

    (a) 100 lbs (c) 20 lbs
    (b) 75 lbs (d) 50 lbs

18. A system having two or more frigerant circuits each with a condenser and evaporator where evaporator of one circuit cools the condenser of the other (lower temp.) circuit, is called a:

    (a) Intercooling system
    (b) Counterflow system
    (c) Cascade system
    (d) Two-step system

19. What are the suggested possibilities of trouble indicated by the gauge and superheat readings; low suction pressure - low superheat:

    (a) Overcharge of refrigerant
    (b) Improper superheat adjustment
    (c) Compressor undersized

SERVICE AND MAINTENANCE QUIZ NO. 9 Section 3

20. A heat pump is a refrigeration system because __________.
    - (a) It is capable of pumping heat as well as cold
    - (b) It can cool below standard air temperature
    - (c) The refrigerant absorbs heat at one temperature and rejects it at a higher temperature.

21. A reverse acting pneumatic humidistat increases the air pressure to a controlled device when the humidity:
    - (a) Increases
    - (b) Decreases
    - (c) Remains static
    - (d) Mixes

22. For a given valve opening, the flow through the valve will increase as the pressure drop __________.
    - (a) Increases
    - (b) Decreases
    - (c) Remains constant
    - (d) Stops

23. The bimetal strip in a thermostat is a __________.
    - (a) The anticipator strip
    - (b) The differential strip
    - (c) The contactor strip
    - (d) None of these

24. On a pneumatic control system the thermostat is connected to the valve or damper motor by a small diameter tubing usually:
    - (a) 1/4" O.D.
    - (b) 3/8" O.D.
    - (c) 3/8" I.D.
    - (d) 1/4" I.D.

25. For a heating application using a normally open valve the controller is:
    - (a) Submaster control
    - (b) Indirect acting run thermostat
    - (c) Direct acting run thermostat
    - (d) Manual switch control

26. What is the size of grille or register to pass 800 cfm at 600 foot velocity? Assume free area at 80%.
    - (a) 133 sq in.
    - (b) 197 sq in.
    - (c) 239 sq in.
    - (d) 314 sq in.

27. Most heat removed from the refrigerant in a condenser is of the __________ variety.
    - (a) Latent heat
    - (b) Specific heat
    - (c) Superheat
    - (d) Overheat

28. A unit of power equal to one horsepower delivered at the shaft of an engine or motor:
    - (a) Electrical horsepower
    - (b) Horsepower
    - (c) Brake horsepower
    - (d) Input power

29. According to American Standard Safety Code for Mechanical Refrigeration, what shall be provided on all systems containing more than 20 pounds of refrigerant:
    - (a) Firestats
    - (b) Pressure limiting devices
    - (c) Receivers
    - (d) Fire extinguishers

30. On a water cooled condenser operating off a cooling tower, approximately how many gallons are required per ton of refrigerant to be circulated through condenser:
    - (a) 3 gallon per minute
    - (b) 2 gallon per minute
    - (c) 4 gallon per minute

31. The unit of pressure commonly used in air ducts is measured in:
    - (a) Inches of water
    - (b) Inches of air
    - (c) Velocity
    - (d) Inches of mercury

SERVICE AND MAINTENANCE QUIZ NO. 9 Section 4

32. Horsepower is what unit of power:

    (a) The effort necessary to raise 36,000 lbs a distance of one foot in one minute

    (b) The effort necessary to raise 33,000 lbs a distance of one foot in one minute

    (c) The effort necessary to raise 24,000 lbs a distance of one foot in one minute

    (d) The effort necessary to raise 12,000 lbs a distance of one foot in one minute

33. The pressure due to the weight of a liquid in a vertical column or more generally the resistance due to lift is called:

    (a) Static head (c) Hydraulic pressure

    (b) Head pressure (d) Temperature head

34. What is the purpose of a thermostat:

    (a) To control the heat

    (b) To control the cold

    (c) To cycle the cooling system

    (d) To control the temperature

35. What is the purpose of the holding circuit in a magnetic motor starter:

    (a) Holds the main contacts open until the control circuit through the interlock is made

    (b) Holds the main contacts closed until the control circuit through the interlock is broken

    (c) Holds the main contacts open until the control circuit through the interlock is broken

    (d) Holds the main contacts closed until the control circuit through the interlock is made

36. Which one of the following is not a common type of steam heating system:

    (a) Vapor (c) Vacuum

    (b) Atmospheric (d) One pipe gravity

37. What type boiler has water around the outside of the tubes being heated by the hot gases within the tubes:

    (a) Water tube boiler (c) Radiant boiler

    (b) Fire tube boiler (d) Atmospheric boiler

38. Where oil fuel tanks are lower than the burner on fuel burning equipment, it is recommended that they have what kind of piping system:

    (a) Plastic pipe system (c) 2 pipe system

    (b) 3 pipe system (d) 1 pipe system

39. Which one of the following is a type of steam heating system:

    (a) Direct expansion (c) Reverse return

    (b) Vacuum (d) Injection

40. Whenever air flows through a pipe or duct, some pressure is lost because of friction, hence the power required to deliver a given quantity of air ________ as the size of the duct is decreased.

    (a) Remains the same (c) Decreases

    (b) Increases

SERVICE AND MAINTENANCE QUIZ NO. 9 Section 5

41. When using a double element time delay fuse with a 20 amp. motor, the maximum fuse size should be:

    (a) 60 amps. (c) 30 amps.

    (b) 20 amps. (d) 25 amps.

42. A two-stage refrigeration system should be used when:

    (a) Ammonia is used as the refrigerant
    (b) Compression ratio would be very high
    (c) Compression ratio would be very low
    (d) Suction temperature would be very high

43. A double riser suction line is used to:

    (a) Allow refrigerant vapor to return more easily
    (b) Cause lower velocity of refrigerant vapor
    (c) Allow the compressor to pump more cfm
    (d) Allow more satisfactory oil return

44. On a motor a running capacitor is used to:

    (a) Increase torque
    (b) Increase horsepower
    (c) Increase power factor
    (d) Increase amperage

45. Fresh meat will keep better when stored in an area with:

    (a) 50% relative humidity
    (b) 65% relative humidity
    (c) 85% relative humidity
    (d) 99% relative humidity

46. The bimetal strip in a thermostat is ________________.

    (a) The anticipator strip
    (b) The differential strip
    (c) The contractor strip
    (d) None of these

47. A heat pump system may meter the refrigerant by using ________________.

    (a) Only a capillary tube
    (b) Only a CP tube
    (c) Only a TX valve
    (d) A capillary tube or TX valve

48. In heat pump terminology, CP is ________________.

    (a) A kind of tube
    (b) Condensing pressure
    (c) Coefficient of performance
    (d) Constant pumping system

49. An auxiliary heat source when used with a heat pump is generally ________________.

    (a) A condenser reheat (c) Electrical heating

    (b) A booster pump (d) Heat of compression

50. If a heat pump is designed with a ground coil heat source it is pumping ________ through the coil.

    (a) H_2O (c) CCL_3F

    (b) CO_2 (d) $CHCLF_2$

SERVICE AND MAINTENANCE QUIZ NO. 10 Section 1

1. When using a double element time delay fuse with a 20 amp. motor the maximum fuse size should be:

   (a) 60 amps. (c) 30 amps.

   (b) 20 amps. (d) 25 amps.

2. On a motor, a running capacitor is used to:

   (a) Increase torque (c) Increase power factor

   (b) Increase horsepower (d) Increase amperage

3. If, upon letting an R-12 refrigeration compressor stand idle long enough for the entire system to cool down to the temperature of the surrounding air of 90°F and the reading of the head pressure gauge is 115 lb., what is this an indication of:

   (a) Overcharge of gas (c) Air in system

   (b) Undercharge of gas (d) Vacuum gauge

4. What is used to determine the amount of moisture remaining in a refrigeration system:

   (a) Vacuum pump (c) Wet bulb indicator

   (b) Vacuum dehydration indicator (d) Vacuum gauge

5. In the discussion of the refrigeration cycle, Compression Ratio is expressed in the following equation thus:

   (a) $\text{Compression Ratio} = \dfrac{\text{Discharge pressure PSI absolute}}{\text{Suction pressure PSI absolute}}$

   (b) $\text{Compression Ratio} = \dfrac{\text{Suction pressure PSI absolute}}{\text{Discharge pressure PSI absolute}}$

   (c) $\text{Compression Ratio} = \dfrac{\text{Discharge temperature absolute}}{\text{Suction temperature absolute}}$

   (d) $\text{Compression Ratio} = \dfrac{\text{Suction temperature absolute}}{\text{Discharge temperature absolute}}$

6. An A.C. single phase motor, split phase, will not start. The cause may be ________________.

   (a) Defective capacitor

   (b) High mica between commutar bars

   (c) Short circuit in the armature windings

   (d) Open circuit in motor winding

7. On an F-12 Refrigeration System, if the suction pressure at the compressor is 35 psi with a 2 psi suction line loss and the suction line temperature taken at point TX valve bulb is fastened, is 51°, what is your Superheat Reading:

   (a) 13°F (c) 11°F

   (b) 14°F (d) 9°F

8. Which of the following electrical wires will carry the most current:

   (a) 14 (c) 12

   (b) 10 (d) 24

9. An electrical network which has 120 volts to a neutral from all legs:

   (a) Delta (c) Polyphase

   (b) Star (d) 2 phase

10. What type of threads are used on flare fittings:

    (a) National fine (c) U.S.A.

    (b) National course (d) Standard pipe threads

11. What fitting should be used to connect 1/4" O.D. copper tubing to a 1/4" internal pipe thread opening in the compressor body:

    (a) A union (c) A tee fitting

    (b) A half-union (d) A street ell

SERVICE AND MAINTENANCE QUIZ NO. 10 Section 2

12. A non-positive displacement compressor:

    (a) Positive (c) Centrifugal

    (b) Rotary (d) Reciprocating

13. Device for removing dust from the air by means of electric charges induced on the dust particles:

    (a) Electric precipitator

    (b) Electric washer

    (c) Electric magnet

    (d) Electric ejector

14. A refrigerant evaporator in which the heat transfer surface is immersed in the refrigerant being evaporated:

    (a) Condenser

    (b) Dry type evaporator

    (c) Flooded evaporator

    (d) TX coil evaporator

15. The gas resulting from the instantaneous evaporation of refrigerant in a pressure-reducing device to cool the refrigerant to the evaporation temperature obtained at the reduced pressure:

    (a) Superheated gas (c) Inert gas

    (b) Flash gas (d) Non-condensible gas

16. Latent heat involved in the change between liquid and vapor states:

    (a) Heat of fusion

    (b) Heat of vaporization

    (c) Heat of reaction

    (d) Heat of the liquid

17. When a dry and wet bulb thermometer are held in a dry air stream the wet bulb will always read lower because ________________.

    (a) A wet bulb thermometer is calibrated lower

    (b) The wet bulb thermometer is cooled by evaporization

    (c) A wet bulb thermometer is always mercury

    (d) It won't read lower

18. When an air sample precipitates moisture, it indicates ________________.

    (a) The leaving air is too warm

    (b) The leaving air is too cold

    (c) The air is not circulating across the coil

    (d) The saturation temperature has been passed

19. A squirrel cage fan is another name for a __________.

    (a) Backward inclined fan

    (b) Forward inclined fan

    (c) A radial flow fan

    (d) An exhauster fan

20. What is common to all filters?

    (a) They must all be coated

    (b) They are all constructed of fiber glass

    (c) They are all 100% fireproof

    (d) They all have a pressure drop

SERVICE AND MAINTENANCE QUIZ NO. 10 Section 3

21. A compressor is short cycling, which of the following would *not* be the cause?

    (a) Low pressure controller differential set too close

    (b) High pressure controller set too close

    (c) A leaky compressor valve

    (d) Improper thermostat operation

22. What alloys shall not be used in contact with any freon refrigerant:

    (a) Aluminum (c) Magnesium

    (b) Zinc (d) Copper

23. A unit of power equal to one-horsepower delivered at the shaft of an engine or motor is commonly known as:

    (a) Amperage (c) Voltage

    (b) Brake horsepower (d) Horsepower

24. What is the range of a thermostat:

    (a) The open and close settings

    (b) The temperature difference

    (c) The capacity of the thermostat

    (d) The variety of the models available

25. Heat being a form of energy cannot be:

    (a) Measured (c) Transferred

    (b) Felt (d) Destroyed

26. Removal of heat from compressed gas between compression stages is known as:

    (a) Recooling (c) Subcooling

    (b) Intercooling (d) Upcooling

27. Ohm's law states ______________.

    (a) $I = \frac{R}{E}$ (c) $R = \frac{I}{E}$

    (b) $R = ER$ (d) $I = \frac{E}{R}$

28. An ampmeter reads 110 volts and 12 amperes, how many kilowatts are being consumed?

    (a) 1.32 (c) 12,000

    (b) 3.21 (d) 120

29. Low suction pressure may be caused by __________.

    (a) Dirty filters

    (b) Restriction in liquid line

    (c) Shortage of refrigerant

    (d) Any of these

30. Low discharge pressure may be caused by __________.

    (a) Dirty filters

    (b) Restriction in liquid line

    (c) Shortage of refrigerant

    (d) None of these

31. An A.C. single phase motor, capator start, will start up but heats rapidly. The cause may be ______________.

    (a) Centrifugal starting switch not opening

    (b) Grounded armature winding

    (c) Open starting winding

    (d) Motor overloaded

SERVICE AND MAINTENANCE QUIZ NO. 10 Section 4

32. An electrical condenser used for power factor correction is known as a :

    (a) Transformer (c) Capacitor

    (b) Starter (d) Relay

33. A ton of refrigeration is a unit of refrigeration capacity corresponding to the removal of how many BTU per day:

    (a) 144,000 Btu's (c) 288,000 Btu's

    (b) 200,000 Btu's (d) 12,000 Btu's

34. A thermal device the opens its contacts when the electric current through a heater coil exceeds the specified value for a given time is called:

    (a) Fuse (c) Control relay

    (b) Relief relay (d) Thermal overload relay

35. What should be first adjustment when adjusting a control which has a cut in differential adjustment:

    (a) Differential

    (b) Gauge for cut-in

    (c) Either differential or range

    (d) Range for cut-out

36. The gas resulting from the instantaneous evaporation of refrigerant in a pressure-reducing device to cool the refrigerant to the evaporation temperature obtained at the reduced pressure:

    (a) Superheat gas (c) Inert gas

    (b) Flash gas (d) Non-condensible gas

37. If three unmarked gas cylinders R-12, R-22 and R-500 respectively are all about one-half full and have been stored in the same room for several days at a temperature of 80°F which cylinder contains the R-22:

    (a) There is no way to tell if not marked

    (b) The cylinder that gives a gauge reading of 103#

    (c) The cylinder that gives a gauge reading of 84#

    (d) The cylinder that gives a gauge reading of 145#

38. Mercury is used in thermometers because it:

    (a) Expands uniformly

    (b) Does not freeze until it reaches-38.2°F

    (c) Does not boil until it reaches 674.6°F

    (d) Does all of the above

39. What will the reading of Wet and Dry Hygrometers be if exposed to air that is completely saturated:

    (a) They will both read alike

    (b) The Wet Bulb will be lower

    (c) The Dry Bulb will be lower

    (d) 90% relative humidity

40. What type of electric motor has a set of field coils and a rotating armature:

    (a) An induction motor

    (b) A capacitor split phase motor

    (c) A repulsion start induction run motor

    (d) Dual winding repulsion induction motor

SERVICE AND MAINTENANCE QUIZ NO. 10 Section 5

41. On a system in operation, a warm liquid line without high head pressure is an indication of:

    (a) Shortage of condenser water

    (b) Expansion valve closed

    (c) Air in condenser

    (d) Shortage of refrigerant

42. In practically all coils used for cooling air, the flow of air and water through the coil are in the ________ direction to each other.

    (a) Water flow (c) Same

    (b) Parallel (d) Opposite

43. On the installation of condensate drains or of well water disposal, it is not permissable to run these pipe lines to a direct connection to a city sewer.

    (a) True (b) False

44. The actual vertical distance through which a condenser or evaporator water pump must lift water is called the:

    (a) Static head (c) Friction head

    (b) Pressure head (d) Vertical head

45. On a pneumatic control system, the main line pressure should be how many psig?

    (a) 25 psig (c) 15 psig

    (b) 5 psig (d) 60 psig

46. A pneumatic control valve which requires air pressure on the bellows or diaphram to close the valve:

    (a) Diverting valve (c) Reverse-acting valve

    (b) Direct-acting valve (d) Normally closed valve

47. A voltmeter is connected into the circuit in which of the following?

    (a) Series (c) Crossover

    (b) Parallel (d) Delta

48. All pressure relief devices (not fusible plugs) shall be:

    (a) Directly pressure-actuated

    (b) Remote pressure-actuated

    (c) Not pressure-actuated

    (d) Manually operated

49. The water valve on a refrigeration system should be connected to the motor electrically in ______.

    (a) Series

    (b) Parallel

    (c) Either series or parallel

    (d) It is not connected to motor

50. What causes electric solenoid valves to chatter?

    (a) Sticking valve

    (b) Valve installed reversed

    (c) Low voltage

    (d) Fluctuating pressure

51. The pressure motor control is usually located on the ____________.

    (a) Condensing unit (c) Cabinet

    (b) Motor (d) Evaporator

52. The low pressure motor control should be connected into the ____________.

    (a) Evaporator (c) Crankcase

GAS LAWS QUIZ NO. 11

1. Figure 1 below shows a cylinder fitted with a freely sliding piston and a pressure gage. The cylinder contains a gas at a pressure of 14.7 psia. The gas in the cylinder occupies 3 cubic feet of space. The temperature is held constant. If the piston is depressed causing the gage to read 13 psig, what is the theoretical volume of the gas? Answer in cubic feet.

   A. 1.575   B. 3.39   C. 5.71   D. 0.1575   E. 15.75

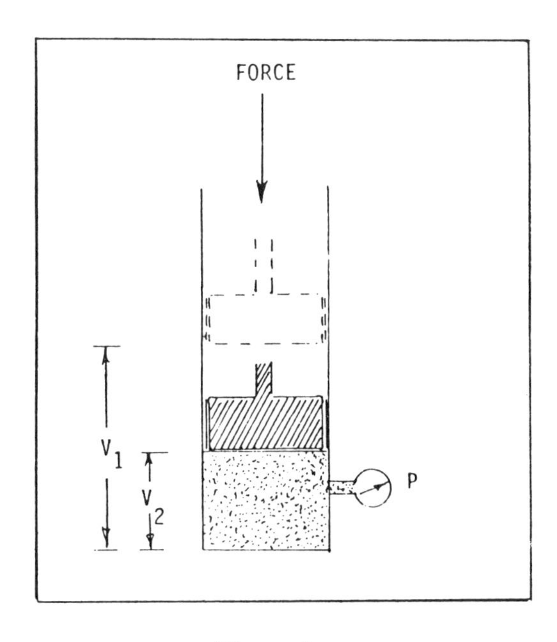

Figure 1

2. A nitrogen pressure of 300 psig is charged in a refrigeration system at 100 F ambient. The nighttime temperature swing is down to 60F. What will the nighttime pressure be?

   A. 277 psia   B. 292 psig   C. 315 psia   D. 180 psig

   E. 277 psig

TEMPERATURE CONVERSION QUIZ NO. 12

Convert the following temperatures from Fahrenheit to Celsius, or Celsius to Fahrenheit.

1. 34 C = ________F

   A. 82.6 B. 93.2 C. 96.4 D. 86.3

2. 32 F = ________C

   A. 32 B. 64 C. 0 D. -32

3. -40 F = ________C

   A. 0 B. -32 C. -16 D. -40

4. 77 C = ________F

   A. 109.6 B. 90.6 C. 137.6 D. 170.6

5. 2004 C = ________F.

   A. 3704 B. 2862 C. 2430 D. 2036

6. 1 C = ________F

   A. 26.2 B. 33.8 C. 32.8 D. -17.2

7. 1 F = ________C

   A. 26.6 B. 33.8 C. -14.7 D. -17.2

8. 82 C = ________F

   A. 27.8 B. 179.6 C. 99.2 D. 148.5

9. 82 F = ________C

   A. 26.6 B. 33.8 C. 27.8 D. 34.3

10. -30 F = ________C

    A. -22 B. -28 C. -32 D. -34

11. -30 C = ________F

    A. -22 B. -28 C. -32 D. -34

12. -140 C = ________F

    A. -220 B. -120 C. -96 D. -84

PIPEFITTING QUIZ NO. 13 Section 1

1. Which valve is most suitable for throttling flow:

   (a) Globe (c) Check

   (b) Gate (d) Swing

2. Flow through swing check valve is in a straight line and without restriction at the seat. This effect on flow is the reason for generally using checks in combination with ____________ valves.

   (a) Globe (c) All

   (b) Lift check (d) Gate

3. Pop safety valve for steam air or gas should always be installed with the stem in a horizontal position.

   (a) True (b) False

4. On the installation of a pressure regulator for use in process work you should always install a relief valve and a pressure gauge on the high pressure side of the regulator.

   (a) True (b) False

5. For the ordinary installation what is considered good rule of thumb practice in spacing at intervals pipe hangers or supports for pipe:

   (a) 5 feet (c) 10 feet

   (b) 20 feet (d) 30 feet

6. In designating the outlets of reduced fittings, whether of the flanged or screwed type, the openings should be read in a certain order. Therefore, on side outlet reducing fittings the size of the side outlet is named in what order:

   (a) First (c) First or last

   (b) According to size (d) Last

7. The seam on any ferrous pipe which is being bent should be located at:

   (a) Top (c) Far side

   (b) Bottom (d) Near side

8. What material may not be used for hanger rods, turnbuckles or clamps:

   (a) Malleable iron (c) Cast iron

   (b) Brass (d) Stainless steel

9. Welding qualification tests on ferrous pipe and materials shall be in accordance with the provisions of:

   (a) American Pressure Vessel Code

   (b) ASME Boiler and Pressure Vessel Code

   (c) Pittsburgh Testing Laboratory

   (d) Hartford Boiler

10. On pressure piping, branch connections made by welding couplings or half couplings directly to the main pipe shall not be used for branches larger than how many inches in nominal pipe size:

    (a) 1" (c) 3"

    (b) 2" (d) 4"

11. On pressure piping screw threads shall conform to:

    (a) 8 threads per inch

    (b) Size of fittings

    (c) U.S. Standard for taper pipe threads

    (d) American Standard for taper pipe threads

PIPEFITTING QUIZ NO. 13 Section 2

12. When making up a pipe joint where you are putting a valve on an 8" long pipe nipple, it is considered good practice to:

    (a) Put the valve in the vise and make up nipple with pipe wrench

    (b) Put the pipe nipple in the vise and make up valve with pipe wrench

    (c) You may either put valve or nipple in vise and make up other with pipe wrench

    (d) Don't use vise to make up this kind of joint, use 2 pipe wrenches

13. What kind of threads does a street ell have:

    (a) External threads only

    (b) Internal threads only

    (c) Both external and internal threads

    (d) NF and NP threads

14. On welded pipe fittings the center to face dimension of a standard 45° 6" weld ell is:

    (a) 3-3/4" (c) 3"

    (b) 5" (d) 6"

15. What fittings should be used to connect 1/4" O.D. copper tubing to a 1/4" internal pipe thread opening in a machine body:

    (a) A union (c) A tee fitting

    (b) A half-union (d) A street ell

16. Find the length of a piece of pipe for a 90° bend with a radius of 40 in. and with two 15 in. tangents.

    (a) 65.82 in. (c) 77.82 in.

    (b) 92.82 in. (d) 62.82 in.

17. When the pitch of the thread on a pipe is 1/4" how many turns are required to thread 2-1/2" of the pipe?

    (a) 8 (c) 12

    (b) 10 (d) 13

18. Two pipes with an area of 3 sq in. and 4 sq in. respectively discharge into a single header. What is the diameter of the header if it has an area = to the sum of the area of the two pipes?

    (a) 4" (c) 3"

    (b) 6" (d) 5"

19. Resistance to flow in a piping system can be decreased most satisfactory by:

    (a) Increasing pipe size

    (b) Increasing the pressure

    (c) Increasing the velocity

    (d) Decreasing the pipe size

20. What fitting should be used to connect 1/4" O.D. copper tubing flared to a 1/4" internal pipe thread opening in a water pump body?

    (a) A union (c) A tee fitting

    (b) A half union (d) A street ell

21. Type K copper tube has a greater wall thickness than type L copper tube.

    (a) True (b) False

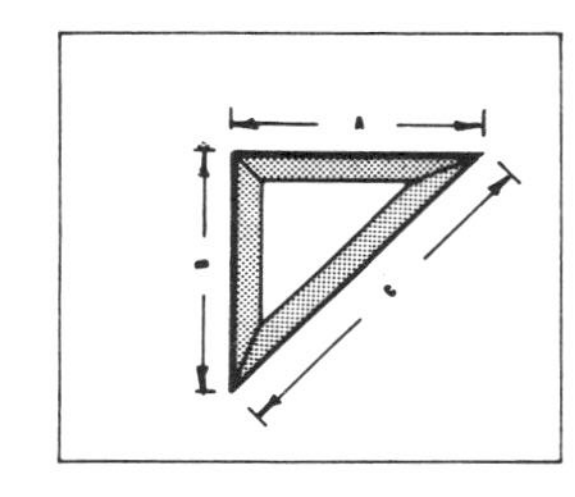

FIGURE 2.7

22. Figure 2.7 shows a 45°angle iron bracket in which the "B" dimension is 24 inches. What is the "C" dimension?

(a) 36 in. (c) 38 in.

(b) 32-1/2 in. (d) 34 in.

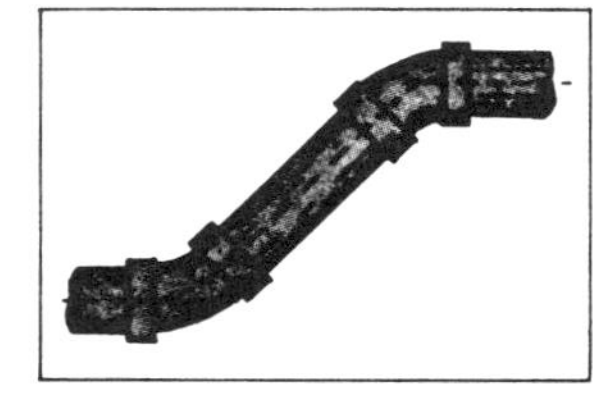

FIGURE 2.8

23. Figure 2.8 shows a 45° offset in which the *travel* is 8-1/2 inches, therefore the *run* is __________.

(a) 10 in. (c) 7 in.

(b) 8-1/2 in. (d) 6 in.

24. A wrap around as used in pipe fitting is used for what purpose:

(a) As a protective device

(b) To draw a straight line around pipe

(c) For insulation

(d) To rig pipe for lifting

25. In pipe bending a general rule for finding the length of any bend is as follows:

(a) R^2 x Angle in Degrees x 0.0175

(b) Radius x Angle in Degrees x 0.0175

(c) R^2 x Angle in Degrees x 3.1416

(d) Radius x Angle in Degrees x 3.1416

26. Which pipe has the thicker wall:

(a) Schedule 20

(b) Schedule 40

(c) Schedule 80

(d) Schedule 30

27. What is the formula for finding the area of a circle:

(a) $A = \pi D^2$ (c) $A = \pi R$

(b) $A = \pi R^2$ (d) $A = \pi D$

28. The plastic flow of pipe within a system is known as:

(a) Corrosion (c) Stress

(b) Strain (d) Creep

PIPEFITTING QUIZ NO. 13 Section 4

29. A straight line drawn from the center to the extreme edge of a circle is known as:

    (a) Circumference of a circle

    (b) Area of a circle

    (c) Diameter of a circle

    (d) Radius of a circle

30. Pipe with its axis in the horizontal plane and rolled during welding so that the weld metal is deposited from the top and within plus or minus, 15 degrees from the vertical place is known as what position weld:

    (a) Pipe-horizontal rooled; Weld-flat position

    (b) Pipe-horizontal fixed; Weld-flat overhead position

    (c) Pipe-vertical fixed; Weld-horizontal position

    (d) Pipe-fixed weld; Vertical or horizontal position

31. The difference between the pressure in a piping system and atmospheric pressure is measured in:

    (a) Pounds per square inch (c) Total pressure

    (b) Absolute pressure (d) Inches of vacuum

32. On industrial gas and air piping, what should be installed on each side of and immediately adjacent to every pressure reducing device:

    (a) A sight glass (c) A valve

    (b) A gauge (d) An orifice plate

33. On industrial gas and air piping, underground cast-iron pipe shall be installed (unless prevented by other underground structures) with a minimum cover of:

    (a) 24" (c) 18"

    (b) 36" (d) 12"

34. A pipe reducer where the center line of the larger pipe is out line with the center of the smaller pipe is known as:

    (a) Concentric reducer (c) Eccentric reducer

    (b) Out of line reducer (d) Bell reducer

35. A pipe fitting shaped like an ell, but with one female end and one male end is called a:

    (a) Male union ell (c) Street ell

    (b) Female union ell (d) Female to male ell

36. Threaded and tapped fittings that are screwed into the end of other fittings or valves to reduce the size of the end openings are known as:

    (a) Reducers (c) Bushings

    (b) Nipples (d) Increasers

37. Which is the proper abbreviations for the fitting as shown in Figure 2.6, the fitting being Standard Black Cast Iron Screwed 125# working pressure:

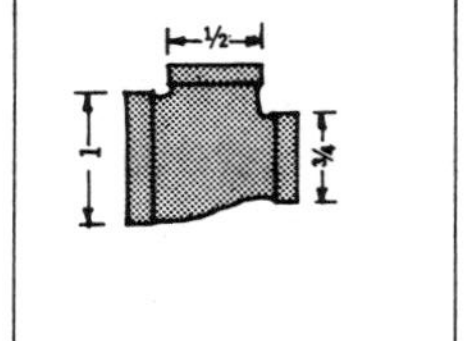

FIGURE 2.6

    (a) 1/2" x 1" x 3/4" std. blk ci scd tee 125# W.P.

    (b) 1" x 1/2" x 3/4" std. blk ci scd tee 125# W.P.

    (c) 3/4" x 1" x 1/2" std. blk ci scd tee 125# W.P.

    (d) 1" x 3/4" x 1/2" std. blk ci scd tee 125# W.P.

REFRIGERATION CYCLE QUIZ NO. 14

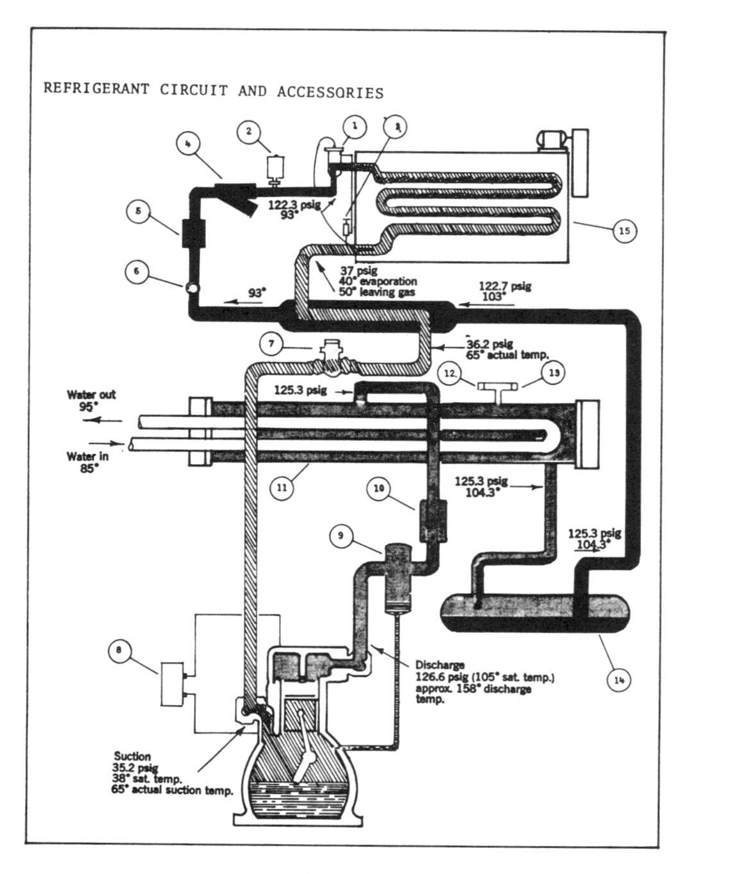

Figure 2

Figure 2, alongside, diagrams a refrigeration circuit with the standard accessories, and shows the change of state between liquid and gas for a system using R-12.

Write the corresponding correct number in for each of the accessories listed.

_______ *Evaporator*

_______ *Expansion valve*

_______ *Evaporator pressure regulator*

_______ *Hi-lo cut-out*

_______ *Solenoid valve*

_______ *Liquid receiver*

_______ *Sight glass*

_______ *Oil separator*

_______ *External equalizer*

_______ *Relief valve*

_______ *Strainer*

_______ *Vent valve*

_______ *Condenser*

_______ *Dehydrator*

_______ *Muffler*

Low pressure gas

Low pressure liquid

High pressure gas

High pressure liquid

Oil

PULLEY LAWS QUIZ NO. 15

1. Given: A 2500 rpm motor is driving a 10 in. fan pulley 5000 rpm.

   Find: Diameter of the motor pulley, inches

   A. 30 B. 15 C. 25 D. 20

2. Given: A 12 in. fan pulley with an rpm of 1250, is driven by a 3 inch motor pulley.

   Find: Rpm of motor pulley

   A. 5000 B. 4000 C. 3750 D. 2750

3. Given: A fan is turning 5000 rpm. The driving motor has a 10 inch pulley at 1000 rpm.

   Find : The diameter of the fan pulley, inches

   A. 2 B. 20 C. 50 D. 5

4. Given: A fan must be driven at 3750 rpm. It has a 4 in. pulley. The motor pulley is 30 inches.

   Find: The speed of the motor, rpm.

   A. 400 B. 500 C. 375 D. 270

5. Given: A 750 rpm fan with a 20 inch pulley, driven by a 4 inch diameter motor pulley.

   Find: The motor rpm.

   A. 3750 B. 2750 C. 1850 D. 1800

FAN LAWS QUIZ NO. 16

1. A 2 hp fan is operating at 1.75" sp at a speed of 1000 rpm and developing 2000 cfm. If the speed is reduced to 800 rpm, what is the new cfm?

   A. 1800 B. 1600 C. 1400 D. 1200

2. What is the new sp?

   A. 1.46 B. 1.12 C. 1.01 D. 0.85

3. What is the new hp?

   A. 1.750 B. 1.256 C. 1.024 D. 0.955

4. A system is found to be 20 percent short of air. Testing shows the fan is actually delivering 4000 cfm. The rpm measures 1000 and the sp is .5. What will the new rpm and sp be?

   A. 1250 rpm and .78 sp C. 1250 rpm and 1.56 sp

   B. 1500 rpm and .78 sp D. 1500 rpm and 1.56 sp

THINGS TO REMEMBER QUIZ NO. 17

1. One gallon of water equals_____pounds.

   A. 16.33 B. 7000 C. 62.37 D. 8.34

2. Absolute temperature equals_____.

   A. °C + 14.7 B. °F + 14.7 C. °F + 460 D. 0°

3. The chemical symbol for Refrigerant 12 is_____.

   A. CCl_2F_2 B. $CHClF_2$ C. CCF_2l_2 D. CHI_2F_2

4. The chemical symbol for Refrigerant 22 is_____.

   A. CCl_2F_2 B. $CHClF_2$ C. CCF_2l_2 D. CHl_2F_2

5. One psi equals_____feet of water.

   A. 0.433 B. 2.31 C. 8.34 D. 14.7

6. Gauge pressure equals_____.

   A. psia - 14.7 C. 0.433 + psia
   B. absolute + 14.7 D. psia + 14.7

7. One cubic foot of water weighs_____pounds.

   A. 8.34 B. 13.34 C. 62.37 D. 14.7

8. One foot of water equals_____.

   A. 62.37 lb B. 7000 grains C. 2.31 psig D. 0.433 psig

9. -40°F =_____°C.

   A. 40 B. 104 C. -32 D. -40

10. Gauge pressure reads 26 psi, what is the feet, head?______.

    A. 11.25 B. 60 C. 74.34 D. 80

FORMULAE QUIZ NO. 18

1. A motor pulley has a pitch diameter of 6 in. Its rpm is 3600. It is driving a fan pulley with a 15 in. pitch diameter. What is the rpm of the fan pulley?

   A. 7080 B. 4200 C. 2460 D. 1440

2. A fan is turning 5000 rpm. The driving motor has a 10 in. pulley at 1000 rpm. Find the pulley pitch diameter.

   A. 2 B. 20 C. 50 D. 5

3. The Empire State Building is 1,250 feet high. If a 40 ft water tank filled with water stood on top of it and a pipe ran down to the ground, what would a pressure gauge at ground level read? All faucets are closed, there is no water drain off.

   A. 541 psi B. 559 psi C. 563 psi D. 2980 psi

4. A 450 pipe offset has a travel of 12 in. What is the offset?

   A. 7.936 B. 8.002 C. 8.204 D. 8.484

5. For a 45° offset a 24 in. "advance" (run), what is the "long-side" (travel)?

   A. 16.97 B. 24 C. 28.24 D. 33.94

6. Figure Z-Z shows a 45° offset around a square obstruction; to find the starting point for the offset, find the distance of A if C = 15 inches and D = 8 inches.

   A. 27 - ½" B. 28 - 9/16" C. 24 - 11/32" D. 26 - 10/32"

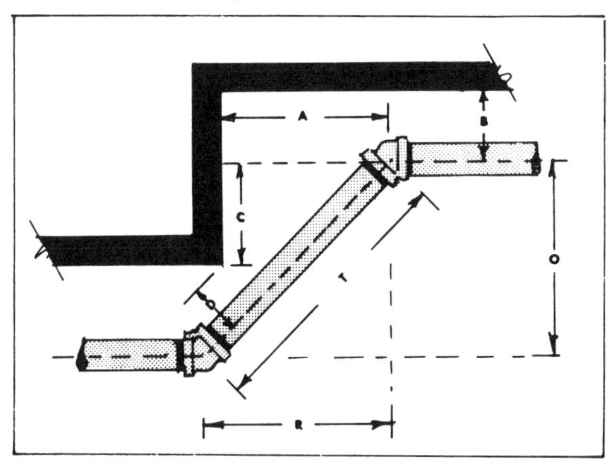

FIGURE Z-Z

QUIZ NO.18 Continued

7. The symbol π stands for the Greek letter pi. It means_____

   A. 4.13 B. 1.34 C. 3.14 D. 14.7

8. A mechanic measures a velocity of 600 fpm discharging from a 30 x 6 wall grille with a 90% free area. Which of the following is the nearest to actual air quantity emerging from the grille.

   A. 600 cfm B. 650 cfm C. 800 cfm D. 815 cfm

9. A return air register is 18" x 24". It is known to be moving 1000 cfm. The free area is 80%. What is the velocity passing through in fpm?

   A. 416 B. 614 C. 518 D. 333

10. One kilowatt equals_____.

    A. 3.415 Btuh B. 3415 Btuh C. .3415 D. 34.15

CODES QUIZ NO.19

1. Joints on copper tubing which are made by the addition of filler metal shall be brazed in refrigerating systems containing Group_____.

   A. I B. II C. III D. II or III

2. Refrigerant 12 and 22 are both classified as Group_____ refrigerants.

   A. I B. II C. III D. II or III

3. Refrigerant R-717 is classified as Group_____refrigerant.

   A. I B. II C. III D. IV

4. Refrigerant R-717 is the number designation for_____.

   A. Sulphur dioxide C. Carbon dioxide
   B. Propane D. Ammonia

5. Refrigerant piping crossing an open space which affords passageway in any building shall not be less than_____ft above the floor.

   A. 6 B. $6\frac{1}{2}$ C. 7 D. $7\frac{1}{4}$

6. Whenever there is an amount of Group 2 refrigerant in a machine room in excess of 110 lb; there must be_____provided at a location close to that machine room.

   A. two masks or helmets
   B. two CO_2 fire extinguishers
   C. four pair of rubber gloves
   D. an automatic alarm bell

7. According to the South Florida Building Code, 4802.1(c), there shall be a complete change of air every three minutes in dry-cleaning plants. If you were replacing an exhaust fan in such a plant whose dimensions were 75 ft x 100 ft x 10 ft high, what size fan (cfm) would you select. Assume no static pressure against the fan.

   A. 10,000 B. 15,000 C. 20,000 D. 25,000

8. The minimum allowable size for air conditioning condensate drain piping is_____inches.

   A. 1/2 B. 3/4 C. 1 D. 1 1/4

9. When PVC piping is installed under a concrete floor slab on fill, it shall be a minimum of 1 1/4 in. diameter, be laid on a firm base and backfilled with 2 in. of sand and be a minimum of _____below the bottom of the slab.

   A. 2 feet B. 12 inches C. 6 inches D. 2 inches

10. Round metal ducts shall be secured against disarrangement at the joint with not less than_____in all buildings other than Group I Occupancy (residential).

    A. a slip drive
    B. A canker clamp
    C. a U-strap
    D. a sheet metal screw

WARM-UP QUIZ NO. 20

1. Electrical resistance is measured in_______.

   A. coulombs B. henrys C. ohms D. Watts

2. To raise 35 pounds of water from 60°F to 70°F requires_____ Btu's.

   A. 20 B. 100 C. 200 D. 350

3. The transferring of heat from one object to anther by means of air current is called_______.

   A. radiation B. convection C. transformation D. conduction

4. The heat picked up in the evaporator must approximately equal_______.

   A. the heat of compression
   B. the heat given up by the condenser
   C. the heat lost by the medium being cooled
   D. the heat gained by the condenser

5. If a drum of dichlorodiflurorometh ane, partially filled with liquid refrigerant is stored in a 100°F room, the pressure in the drum would be_______.

   A. 132 psig B. 117 psig C. 32"HG D. 196 psig

6. How much condenser water is required for a five ton air conditioning system using a cooling tower with a 10° rise?

   A. 14 gpm B. 10 gpm C. 30 gpm D. 8 gpm

7. Refrigerant lines crossing a passageway must leave a vertical clearance of______, or be against the ceiling.

   A. 6'6' B. 7'0" C. 7'3" D. 8'6"

8. Stop valves used with soft annealed copper tubing or hard drawn copper tubing_______inches outside diameter or smaller shall be securely mounted, independent of tubing fastenings or supports.

   A. 3/8 B. 1/2 C. 5/8 D. 7/8

9. The volume of one pound of air 75°F and normal atmospheric pressure is approximately_______cubic feet.

   A. 10.7 B. 13.7 C. 15.7 D. 17.7

10. Referring to OSHA'S Standards and interpretations, the word "shall" means_______.

    A. recommended C. the same as the word "should"
    B. mandatory D. none of the above

FINAL EXAMINATION
FOR
AIR CONDITIONING JOURNEYMAN

This is your final exam. You may expect something similar to this in the actual county examination for journeyman air conditioning.

Depending upon which county is giving the examination it may be open or closed book--or both. It may be a 3-hour or 6- hour exam. A grade of 75% is required to pass most exams.

The following examination is a 6-hour open and closed book exam. It is important that you take this test under conditions that simulate the actual exam room. If you cannot have quiet, uninterrupted space and time at home, go to your local library.

The morning session is closed book. Do not refer to any book while taking this exam. You are allowed three hours. The afternoon session is open book. You may use any books you wish as reference for this test.

Grade yourself upon completion. You need 75% to pass. This will give you a good idea of how ready you are. Study those areas in which you are weak.

The quality of typing and reproduction you are looking at in this book is considerably better than you will find in most examination rooms for tradesmen, either county, municiple or state. Be prepared for poor graphic work, poor illumination, and discomfort when you enter the actual exam room.

In the exam room you will usually be required to mark your answers on a machine grading score card. On the next page you may see and example of such a card. In the exam room, it will be wise to first mark your answers on the actual test paper and then transfer to the machine grading score card. Try not to erase on the score card.

AMERICAN COMMUNITY SERVICES, INC.
Testing and Measurement Division

NAME ______________________________
print (last) (first) (middle)

ADDRESS ______________________________

____

LOCATION ______________ SIGNATURE ______________

DIRECTIONS FOR MARKING

- Use black lead pencil only (No. 2½ or softer).
- Do NOT use ink or ballpoint pens.
- Make heavy black marks that fill circle completely.
- Erase cleanly any answer you wish to change.
- Make no stray marks on this answer sheet.

TEST NUMBER	I.D. NUMBER				
3	0	0	0	2	5
0	0	0	0	0	0
1	1	1	1	1	1
2	2	2	2	2	2
3	3	3	3	3	3
4	4	4	4	4	4
5	5	5	5	5	5
6	6	6	6	6	6
7	7	7	7	7	7
8	8	8	8	8	8
9	9	9	9	9	9

1 A B C D E	35 A B C D E	69 A B C D E	103 A B C D E	137 A B C D E	171 A B C D E
2 A B C D E	36 A B C D E	70 A B C D E	104 A B C D E	138 A B C D E	172 A B C D E
3 A B C D E	37 A B C D E	71 A B C D E	105 A B C D E	139 A B C D E	173 A B C D E
4 A B C D E	38 A B C D E	72 A B C D E	106 A B C D E	140 A B C D E	174 A B C D E
5 A B C D E	39 A B C D E	73 A B C D E	107 A B C D E	141 A B C D E	175 A B C D E
6 A B C D E	40 A B C D E	74 A B C D E	108 A B C D E	142 A B C D E	176 A B C D E
7 A B C D E	41 A B C D E	75 A B C D E	109 A B C D E	143 A B C D E	177 A B C D E
8 A B C D E	42 A B C D E	76 A B C D E	110 A B C D E	144 A B C D E	178 A B C D E
9 A B C D E	43 A B C D E	77 A B C D E	111 A B C D E	145 A B C D E	179 A B C D E
10 A B C D E	44 A B C D E	78 A B C D E	112 A B C D E	146 A B C D E	180 A B C D E
11 A B C D E	45 A B C D E	79 A B C D E	113 A B C D E	147 A B C D E	181 A B C D E
12 A B C D E	46 A B C D E	80 A B C D E	114 A B C D E	148 A B C D E	182 A B C D E
13 A B C D E	47 A B C D E	81 A B C D E	115 A B C D E	149 A B C D E	183 A B C D E
14 A B C D E	48 A B C D E	82 A B C D E	116 A B C D E	150 A B C D E	184 A B C D E
15 A B C D E	49 A B C D E	83 A B C D E	117 A B C D E	151 A B C D E	185 A B C D E
16 A B C D E	50 A B C D E	84 A B C D E	118 A B C D E	152 A B C D E	186 A B C D E
17 A B C D E	51 A B C D E	85 A B C D E	119 A B C D E	153 A B C D E	187 A B C D E
18 A B C D E	52 A B C D E	86 A B C D E	120 A B C D E	154 A B C D E	188 A B C D E
19 A B C D E	53 A B C D E	87 A B C D E	121 A B C D E	155 A B C D E	189 A B C D E
20 A B C D E	54 A B C D E	88 A B C D E	122 A B C D E	156 A B C D E	190 A B C D E
21 A B C D E	55 A B C D E	89 A B C D E	123 A B C D E	157 A B C D E	191 A B C D E
22 A B C D E	56 A B C D E	90 A B C D E	124 A B C D E	158 A B C D E	192 A B C D E
23 A B C D E	57 A B C D E	91 A B C D E	125 A B C D E	159 A B C D E	193 A B C D E
24 A B C D E	58 A B C D E	92 A B C D E	126 A B C D E	160 A B C D E	194 A B C D E
25 A B C D E	59 A B C D E	93 A B C D E	127 A B C D E	161 A B C D E	195 A B C D E
26 A B C D E	60 A B C D E	94 A B C D E	128 A B C D E	162 A B C D E	196 A B C D E
27 A B C D E	61 A B C D E	95 A B C D E	129 A B C D E	163 A B C D E	197 A B C D E
28 A B C D E	62 A B C D E	96 A B C D E	130 A B C D E	164 A B C D E	198 A B C D E
29 A B C D E	63 A B C D E	97 A B C D E	131 A B C D E	165 A B C D E	199 A B C D E
30 A B C D E	64 A B C D E	98 A B C D E	132 A B C D E	166 A B C D E	200 A B C D E
31 A B C D E	65 A B C D E	99 A B C D E	133 A B C D E	167 A B C D E	201 A B C D E
32 A B C D E	66 A B C D E	100 A B C D E	134 A B C D E	168 A B C D E	202 A B C D E
33 A B C D E	67 A B C D E	101 A B C D E	135 A B C D E	169 A B C D E	203 A B C D E
34 A B C D E	68 A B C D E	102 A B C D E	136 A B C D E	170 A B C D E	204 A B C D E

FINAL EXAMINATION

MORNING SESSION

You are allowed three hours to complete this test. This is a closed book exam--no references allowed.

Answers to this exam will be found at the end of this book.

THIS EXAM CONSISTS OF FIVE PARTS:

Part 1	Duct Symbols Ten questions each valued at 2 points	20 points
Part 2	Field Wiring Symbols Ten questions each valued at 2 points	20 points
Part 3	Wiring Diagram Problem Complete the schematic	20 points
Part 4	Code Problems Ten questions each valued at 2 points	20 points
Part 5	Refrigerating System Classification Six questions each valued at 3.33 points	20 points

DUCT SYMBOLS (20 points)

Identify the common duct symbols below by placing the correct number alongside of the matching symbol. Select a number from the list below.

A. 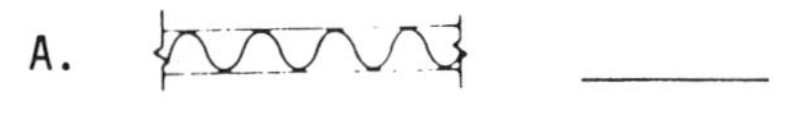 ______

F. ______

B. 

______

G. ______

C. ______

H. ______

D. ______

I. 12/20 ______

E. 20/12 ______

J. 20/12 ______

1. Back draft damper
2. Flexible connection
3. Manual volume damper
4. Supply outlet
5. Side wall register
6. Return air grille
7. Undercut door
8. Acoustical lining
9. Transition piece
10. Cowl and flashing
11. Flexible sound absorber
12. Mixing box
13. Turning vanes
14. Flexible duct
15. Adjustable blank off
16. Duct section, positive pressure
17. Change of elevations
18. Direction of flow
19. Duct size
20. Duct section, negative pressur
21. Elbowed chimney
22. Resistance damper
23. Grille size

FIELD WIRING SYMBOLS (20 points)

Identify the common electrical symbols below by placing the correct number alongside of the matching symbol. Select a number from the list below.

A. ______

B. ______

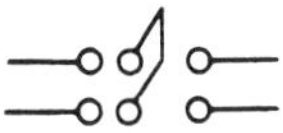

C. ______

D. ______

E. ______

F. ______

G. ______

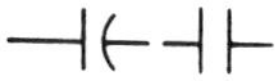

H. ______

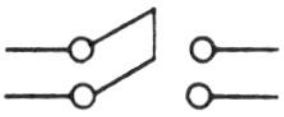

I. ______

J. ______

1. Strip heater
2. Normally open contacts
3. Single pole circuit breaker
4. Overload heater
5. Normally closed contactor
6. Double pole, double throw switch
7. Fuse
8. Hot leg connection
9. Ground connection
10. Switch, double pole, single throw
11. Contacts, normally closed
12. Relay coil
13. Magnetic core coil
14. Adjustable heat anticipator
15. Normally open push switch
16. Capacitor
17. Connected wires

WIRING DIAGRAM PROBLEM (20 points)

Figure 3 shows a typical schematic wiring diagram for an air cooled heat pump with three steps of electric strip heaters for supplementary heating when the unit is in the reversing defrost mode. The field wiring is represented by a broken line symbol.

In this case, the journeyman mechanic who was doing the field wiring assembly (24-V circuit) was suddenly taken ill before completing his work. You are called upon to complete his work. Your partner tells you, "I have completed everything except the transformer hot side, the fan circuit, and the strip heater relays."

Complete the circuit by carefully and neatly drawing in the broken lines where necessary. Neatness will count in your grade.

REFRIGERANT SYSTEM CLASSIFICATION PROBLEM (20 points)

Figure 4 shows a schematic for six different types of refrigerating systems according to ASHRAE 15-1989. Match the system numbers to the classifications below.

A. Indirect vented closed surface system

B. Direct system

C. Indirect open spray system

D. Double direct system

E. Double indirect vented open spray system

F. Indirect closed surface system

FIGURE 3

WITH ACCESSORY MODULATING DAMPER CONTROL, ELECTRIC RESISTANCE HEATERS AND EMERGENCY HEAT SWITCH

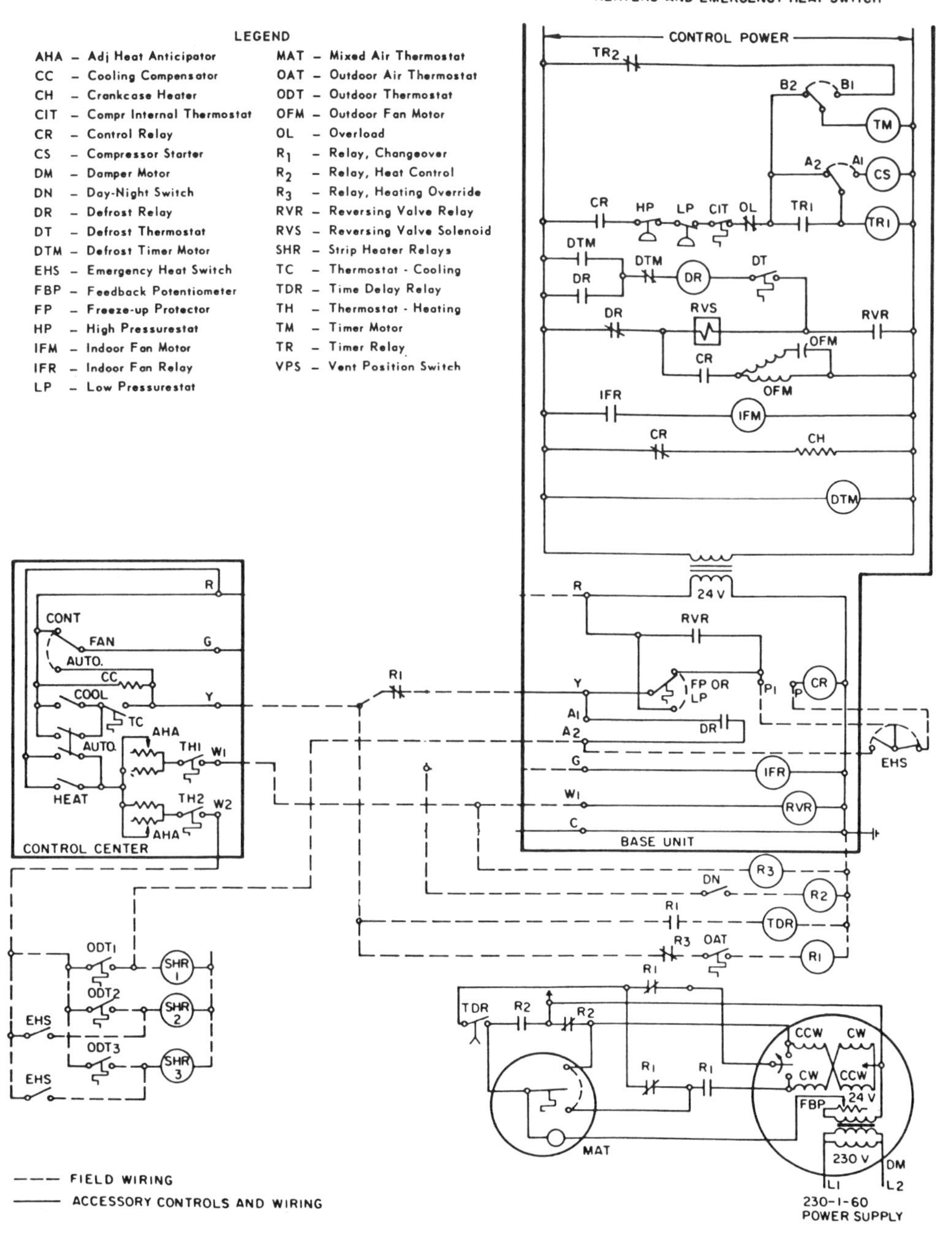

CODE PROBLEMS (20 points)

Answer True or False

1. The minimum size condensate for any air unit drain shall be one inch.

   T_____ F_____

2. A toilet room with air conditioned ventilation air is 10' x 14' x 8'. It therefore requires at least 149.33 cfm exhaust.

   T_____ F_____

3. If a major repair on an air conditioning system does not change the location, size, or capacity of a compressor, coil or duct, no permit is required.

   T_____ F_____

4. Outside air intake must be a minimum of five feet distance from any vent terminal of a plumbing system sanitary line.

5. When an attic fan is used to ventilate a living area it shall have a fire damper shut off operated by a fire stat.

   T_____ F_____

6. The length and width of commercial kitchen hoods shall extend a minimum of 12 inches beyond the appliance over which they are installed.

   T_____ F_____

7. In paint spray areas and lay up areas of fiberglass boat manufacturing places and similar hazardous locations, there shall be a complete change of air every minute.

   T_____ F_____

8. Refrigerant piping crossing an open space which affords passageway in any building shall be not less than 7½ ft (2.29 m) above the floor unless against the ceiling of such space.

   T_____ F_____

9. Commercial kitchen hoods shall not be raised more than 7½ feet above the floor.

   T_____ F_____

10. Every refrigerating system above 6 pounds shall be protected by a pressure-relief device or some other means designed to safely relieve pressure due to fire or other abnormal conditions. Systems under 6 pounds are exempt

    T_____ F_____

FIGURE 4

REFRIGERATING SYSTEM CLASSIFICATION PROBLEM (20 points)

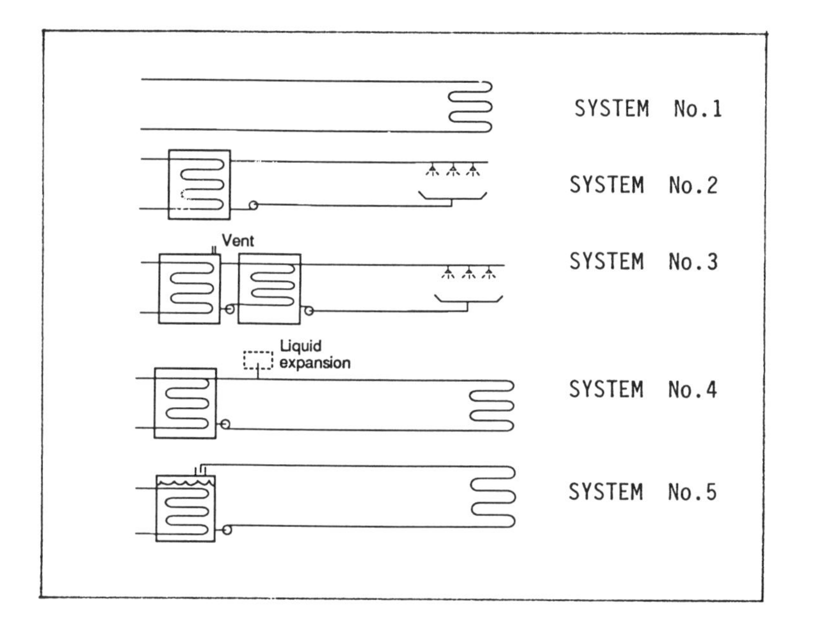

FINAL EXAMINATION

AFTERNOON SESSION

You are allowed three hours to complete this test. This is an open book exam--you may refer to any reference books you wish, for example,

1. Modern Refrigeration & Air Conditioning, Althouse, Turnquist, Bracciano
2. Trane Reciprocating Refrigeration Manual
3. ASHRAE Standard 15-1978, Safety Code for Mechanical Refrigeration
4. Your local Code Book.
5. The Exam Book for Air Conditioning Journeyman, J. Gladstone
6. Journeyman General Mechanical Examination, J. Gladstone
7. Air Conditioning and Mechanical Trades, J. Gladstone
8. The NFPA Fire Codes, 90 90A, 91, 96
9. The SMACNA Manuals; Fiberglass, Low Pressure Ducts
10. The Pipefitters Handbook, Lindsey
11. Red Cross First Aid pamphlets for burns and CPR

THIS EXAM CONSISTS OF FOUR PARTS:

Part 1	First Aid Five questions each valued at 2 points	10 points
Part 2	General Questions Fifty questions each valued at 1 point	50 points
Part 3	First Aid Five questions each valued at 2 points	10 points
Part 4	Steam Specialties Identification Fifteen questions each valued at 2 points	30 points

t 1 FIRST AID (10 points)

A person is suffering from strangulation; therefore, you may have to ______________.

1. Place the victim in a seated position and push his head forward, gently, as far as it will go. T_____ F_____

2. Look into the mouth to see if there is an obstruction, stick your finger down and remove it. T_____ F_____

3. If the victim is lying across an electrical wire, pull him clear immediately. T_____ F_____

4. Deliver sharp blows with the heel of your hand on the victin's spine, between the shoulder blades. T_____ F_____

5. Proceėd with mouth-to-mouth ventilation. T_____ F______

t 2 GENERAL QUESTIONS (50 points)

6. To charge a system with vapor it must be connected to the suction line.

   T_____ F_____

7. To charge a system with liquid it must be connected to the suction line.

   T______F______

8. A capillary tube system is____________.

   A. more accurate C. simpler
   B. variable D. better for large systems

9. A rise in temperature at the condenser will result in___________.

   A. blowing out the gas C. formation of flash gas
   B. better condensing D. more sub-cooling at the TXV

10. The oil separator must be located_________.

   A. lowest point C. between condenser and evaporator
   B. highest point D. between the compressor and condenser

11. The T-X-valve is used to__________.

   A. control the evaporator superheat C. control the gas
   B. control the sub cooling D. none of the above

12. A compound gauge reading on the low side will read the vacuum in__________.

   A. inches of water C. psia
   B. psig D. inches of mercury

13. The frost line method of charging__________.

    A. Can only be used on an open system
    B. Requires a reduction in evaporator load
    C. Should produce no suction line frosting
    D. Requires a manually operated watering device.

14. Before charging a system how long should the system be held under vacumm __________?

    A. 15 min   C. 24 hrs
    B. 2 hrs   D. 48 hrs

15. A split phase motor has__________.

    A. a separate winding for starting
    B. two separate windings for starting
    C. a separate winding for running
    D. A & C
    E. C & B

16. Suction line frosting is caused by__________.

    A.
    B.
    C.
    D.

17. A refrigeration system should be evacuated to________inches Hg before charging.

    A. 14.7   C. 28
    B. 17   D. 38

18. The nominal thickness of galvanized sheet metal for duct systems for 30 gauge is__________inches

    A. .0157   B. .0172   C. .0187   D. .0202

19. A relay converts electrical energy to__________emergy

    A. heat   C. chemical
    B. mechanical   D. transmittable

20. A fiberglass ductboard 60" x 1" is formed into a square duct what is the inside dimension?___________

    A. 16 x 16   C. 14 x 14
    B. 15 x 15   D. 13 x 13

21. Before disconnecting a capacitor it must be____________.

    A. dried   C. sparked on open
    B. discharged by shorting   D. unplugged

. When the static pressure is increased the velocity is_________

A. decreased
B. increased
C. sublimated
D. unchanged

. Which gauge electrical wire has the lowest carrying capacity?

A. 24 B. 22 C. 12 D. 10

. What is the minimum allowable width of pressure sensitive duct tape?

A. 2 in. B. 2½ in. C. 2 3/4 in. D. 3 in.

. The required hanger spacing for a run of 70" ID fiberglass duct is_________ ft O.C.

A. 2 B. 3 C. 2½ D. 4

. The tool one must use to make a male shiplap cut in a fibrous duct layout is a_________tool.

A. green
B. purple
C. blue
D. mitre knife

. The tool one must use to make a rabbet on the end of the board (end cutoff) is a_____________tool.

A. orange
B. blue
C. sharp square-knife
D. red

. Never use acetylene to develop pressure when testing a system for leaks because_________.

A. It is not a sensitive enough gas
B. It will decompose and explode over 30 psi
C. It is flammable
D. It is toxic

. Never use oxygen to develop pressure when testing for leaks because_________.

A. Oxygen is not miscible with refrigerants
B. Oxygen will stratify in the presence of any refrigerant
C. Oxygen will decompose before the test can be made
D. Oxygen will explode in the presence of oil

. When testing for leaks it is first necessary to establish a positive pressure in the system of_________psi.

A. 5 to 30
B. 25 to 50
C. 55 to 75
D. None of the above

31. A reading of 1" WG is equivalent to________.

A. 1 inch psi   B. 2½ psf   C. 5.2 psf   D. 3.5 psi

32. Dirty filters are more efficient dust removers than clean filters________

A. True   B. False

33. When using nitrogen within the pipe during brazing the pressure of the nitrogen should be_____________.

A. very low
B. 50 to 100 psig
C. 250 to 300 psig
D. slightly below atmospheric

34. With unloading compressors, the suction lines should be taken from the __________of the manifold.

A. top
B. bottom
C. center
D. side

35. A partial coil indicates a condition of__________superheat.

A. no   B. low   C. high   D. medium

36. To measure in thousandts of an inch which of the following instruments must be used?

A. depth gauge
B. psychrometer
C. manometer
D. micrometer

37. In a direct expansion chiller the refrigerant is on the________of the tube

A. Inside
B. Outside
C. Both sides
D. Overflow

38. A capacitor motor has__________.

A. a double-start winding
B. 120 cycle current
C. a double-run winding
D. a centrifugal switch

39. A heat exchanger in the liquid line will_________.

A. subcool the liquid
B. Reduce flash gas in the liquid line
C. Reduce liquid refrigerant in the suction line
D. All of the above

40. The oil pressure safety switch responds to the_________

    A. difference between the pump discharge pressure & the suction pressure
    B. Crankcase pressure minus the condenser pressure
    C. Pressure in the discharge valve plates
    D. High compression ratio.

41. The recommended minimum design testing pressures for R-12 according to Safety Code installations are________low side and _________high side, for an air cooled system.

    A. 85 - 127 C. 85 - 169
    B. 144 - 169 D. 127 - 278

42. When pressure testing with either carbon dioxide or nitrogen you must always take care to_______.

    A. Use goggles
    B. Keep the clylinder submerged in water
    C. Use a pressure regulator
    D. All of the above

43. Never go over_________psi pressure when testing for leaks with nitrogen or carbon dioxide.

    A. 120 B. 170 C. 190 D. 210

44. Refrigerant piping crossing an open space which affords passageway in any building shall be not less than 7½ (ft) above the floor unless against the ceiling of such space. __________

    A. True B. False

45. The maximum permissable quantity of R-12 in lbs per 1000 cu ft of room volume shall be__________.

    A. 31 B. 23 C. 16 D. None of the above

46. Which group of refrigerants are considered most hazardous__________.

    A. I B. II C. III D. IV

47. A system using a brine cooled by the refrigerant is considered a_________.

    A. secondary C. reverse return
    B. direct D. indirect

48. Systems using ammonia as the refrigerant are usually piped in_________

    A. copper C. iron or steel
    B. copper or brass D. mechanical joints

49. Electrical consumption is measured in________.

A. columbs B. current C. watts D. volts

50. The pressure gauge permanently connected to the high side of the system must be calibrated______times the design pressure.

A. 2½ B. 1.2 C. 2 D. 10

51. When PVC condensate drains are piped under concrete floor slabs, the top of the pipe shall be a minimum of 2½ inches below the bottom of the slab.

T________ F________

52. Ducts located under concrete slabs shll have at least 2 inches of concrete all around.

T________ F________

53. Substitution of kind of refrigerant in a system shall not be made witho the permission of the________.

A. approving authority
B. user
C. manufacturer
D. all of the above

54. Where Group II refrigerants are used in excess of 110 lb, at least two masks or helmets shall be provided convenient to the machinery room.

T________ F________

55. The duct clearance from metal ducts to combustible construction, includ plaster on wood lath, shall be not less than 1/2 inch.

T________ F________

Part 3 FIRST AID (10 points)

1. For a chemical burn in the eye, wash the eye with running water for at least_______minutes

A. 5 B. 10 C. 15 D. 20

2. For a chemical burn on the skin, wash the chemical off for at least____ minutes.

A. 5 B. 10 C. 15 D. 20

3. If charred clothing is sticking to a burn you should bandage directly ov the clothing.

T________ F________

4. Always add an anti-bacterial agent to the water before soaking a burn.

T________ F________

5. In case of a large, thin burn, first wash it with water.

T________ F________

Part 5 STEAM SPECIALTIES (30 points)

Identify each of the following steam specialties by assigning the proper letter--from A to L--to each of the 15 symbols. Exam room questions with illustrations are usually made from photocopy material and will look very much like these; difficult to read.

A. Steam trap
B. Float and thermostatic steam trap
C. Inverted bucket steam trap
D. Balancing fitting
E. Balanced pressure thermostatic steam trap
F. Balancing valve
G. Globe valve
H. Liquid expansion thermostatic steam valve
I. Gate valve
J. Scraper strainer
K. Pump
L. Air vent

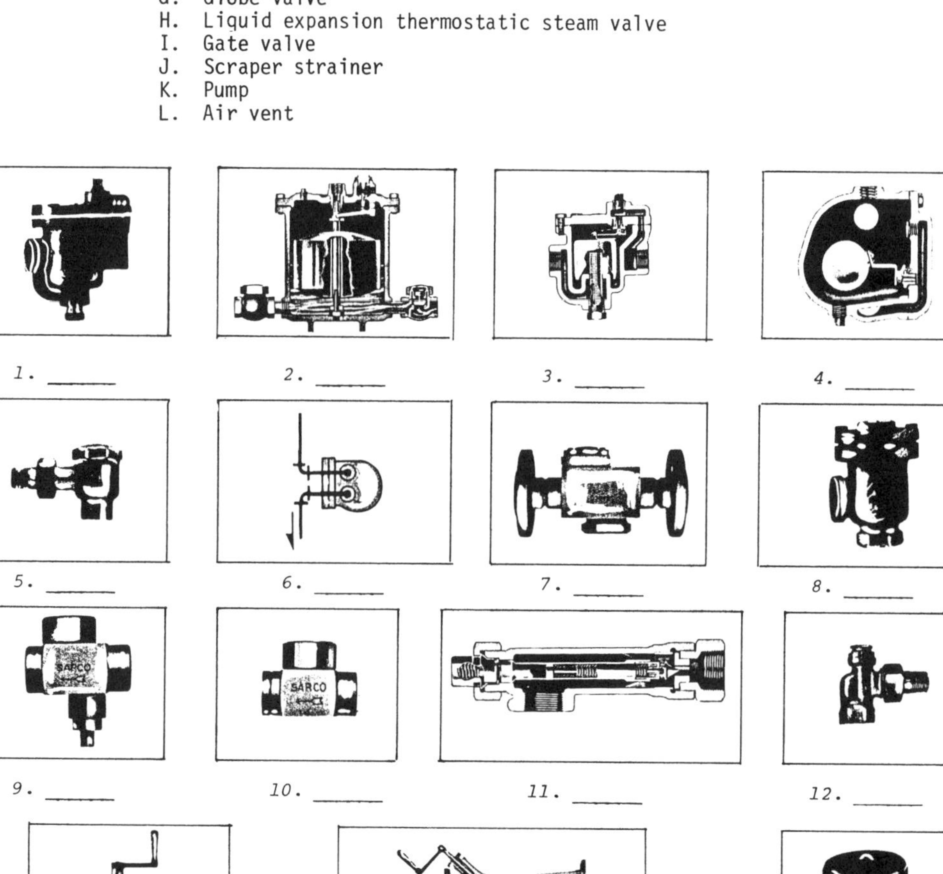

1. ______ 2. ______ 3. ______ 4. ______

5. ______ 6. ______ 7. ______ 8. ______

9. ______ 10. ______ 11. ______ 12. ______

13. ______ 14. ______ 15. ______

PART 5

ANSWERS TO QUIZZES

ANSWER SHEET
300 QUESTIONS

1-B	51-B	101-B	151-C	201-B	251-B
2-B	52-B	102-A	152-D	202-A	252-B
3-A	53-D	103-C	153-A	203-B	253-B
4-A	54-A	104-D	154-C	204-C	254-B
5-A	55-D	105-B	155-B	205-B	255-B
6-B	56-A	106-A	156-B	206-C	256-B
7-A	57-C	107-B	157-C	207-C	257-B
8-D	58-C	108-C	158-C	208-A	258-C
9-C	59-A	109-C	159-A	209-C	259-C
10-D	60-A	110-A	160-A	210-A	260-C
11-B	61-A	111-C	161-C	211-B	261-B
12-B	62-D	112-A	162-C	212-D	262-C
13-A	63-A	113-A	163-A	213-D	263-C
14-D	64-B	114-A	164-B	214-B	264-B
15-D	65-D	115-B	165-B	215-C	265-B
16-C	66-C	116-A	166-A	216-B	266-A
17-A	67-A	117-B	167-A	217-C	267-B
18-C	68-C	118-B	168-A	218-C	268-A
19-C	69-C	119-A	169-A	219-D	269-D
20-D	70-A	120-B	170-A	220-C	270-B
21-A	71-D	121-A	171-B	221-D	271-C
22-B	72-A	122-B	172-C	222-D	272-B
23-D	73-C	123-C	173-C	223-C	273-C
24-B	74-C	124-A	174-C	224-A	274-B
25-B	75-D	125-B	175-B	225-B	275-C
26-A	76-C	126-A	176-B	226-D	276-A
27-D	77-A	127-B	177-C	227-C	277-B
28-A	78-B	128-A	178-A	228-B	278-C
29-B	79-B	129-A	179-A	229-B	279-B
30-C	80-C	130-C	180-B	230-C	280-A
31-C	81-C	131-C	181-C	231-D	281-C
32-B	82-D	132-A	182-A	232-C	282-B
33-B	83-D	133-B	183-A	233-B	283-A
34-D	84-B	134-B	184-B	234-A	284-C
35-C	85-B	135-B	185-C	235-B	285-A
36-A	86-B	136-D	186-D	236-D	286-D
37-B	87-B	137-B	187-B	237-A	287-C
38-D	88-C	138-D	188-C	238-C	288-A
39-A	89-A	139-C	189-C	239-D	289-A
40-D	90-B	140-B	190-D	240-B	290-C
41-D	91-A	141-B	191-B	241-A	291-B
42-C	92-A	142-C	192-A	242-A	292-B
43-B	93-A	143-B	193-A	243-A	293-D
44-C	94-A	144-B	194-D	244-B	294-B
45-A	95-B	145-B	195-A	245-A	295-C
46-C	96-A	146-D	196-D	246-C	296-A
47-B	97-D	147-A	197-D	247-B	297-D
48-A	98-C	148-D	198-C	248-A	298-B
49-C	99-C	149-C	199-D	249-A	299-C
50-D	100-B	150-A	200-B	250-D	300-C

ANSWERS TO QUIZZES

QUIZ NO. 1 Basic Refrigeration

1-B	3-C	5-A	7-B	9-B
2-A	4-A	6-D	8-B	10-B

QUIZ NO. 2 Basic Refrigeration

1-C From Trane AC Manual, Steam Tables p 358 go to column 8, find 1150.4 Go to column 6, find 18.07. Subtract 1150.4-18.07 = 1132.33
2-D From column 6 at 212F, find 180.7. From column 6 at 50F, find 18.07 Subtract 180.07-18.07 = 162 Btu sensible, then, 162/1132 = 14% sensi
3-A Subtract 14% from 100% = 86%
4-A Enter column 2, 14.696 psia and read 212F in column 1.
5-E
6-C
7-E
8-D From Trane AC Manual, p 397, table 6-1, enter Column 3 at 206.62 and read 140F from column 1.
9-C From column 8 across, read 89.967.
10-B From any pressure-temperature chart the difference between the high side and low side (105F and 45F) will be close to 3 times.

QUIZ NO. 3 Cooling Tower

Static discharge head (lift) minus static suction head (fall) equals total static head. The static head of a cooling tower is the vertic distance in feet between the free water level of the cooling tower p and the point of actual water discharge at the tower spray.

50' - 40' = 10'   10' x .433 = 4.33 psi   ANSWER

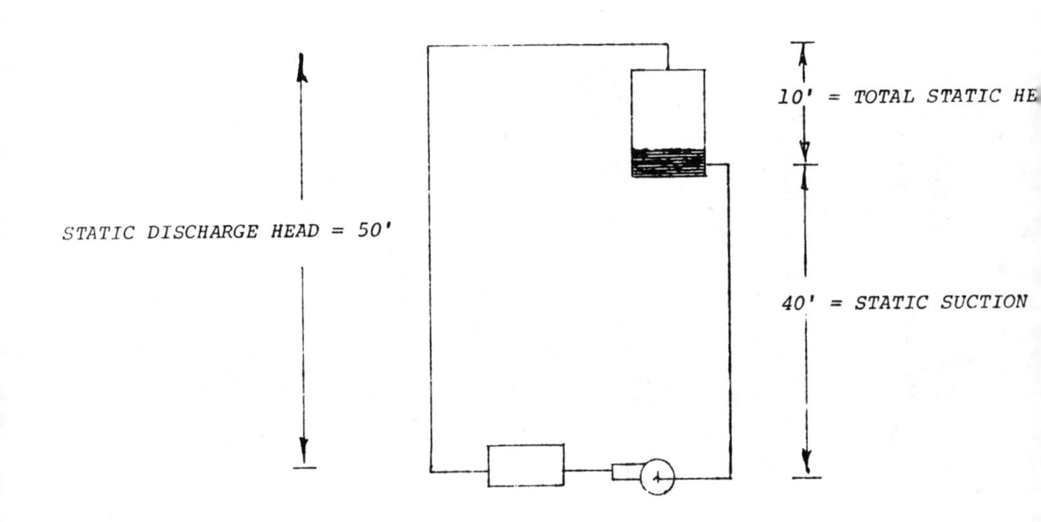

QUIZ NO. 4 Cooling Tower

260' x .433 = 112.58 psig

Answer, A

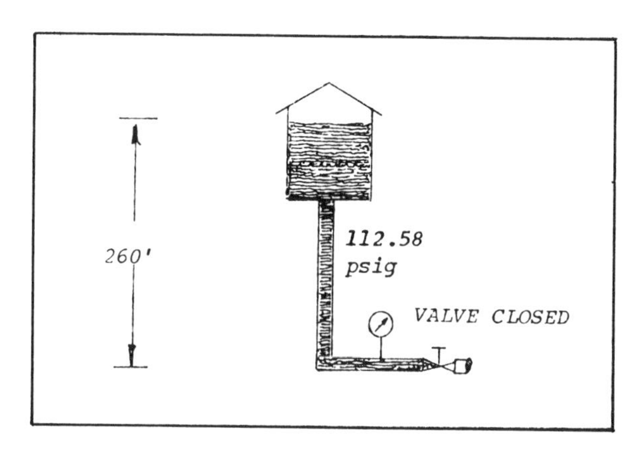

QUIZ NO. 5 Cooling Tower

Gauge No. 1 reads 25.98 psi: 25.98 x 2.31 = 60' H_2O

Gauge No. 2 reads 17.32 psi: 17.32 x 2.31 = 40' H_2O

The differential between the discharge and suction gauges indicates a head of 60-40 = 20 ft H_2O. However, the difference in the pysical height between the two gauges must be taken into account;

$$20 \text{ ft} + 1.5 \text{ ft} = 21.5 \text{ ft } H_2O$$

If the elevated gauge was on the suction side, the answer would have been:

$$20 \text{ ft} - 1.5 \text{ ft} = 18.5 \text{ ft } H_2O$$

ANSWER SHEET

MATH QUIZ NO. 6

1. 251.05 (+) 33.582 (+) 6953.01 (+) 52 (+) .03 (=) 7289.672
2. 953.115 (+) 66.05 (-) 57.32 (+) 63.1 (-) 362.132 (=) 662.813
3. (-) 2 (+) (-) 8 (+) (-) 6 (=) -16
4. (-) 8 (+) 6 (+) (-) 7 (+) 4 (+) 16 (+) (-) 2 (=) 9
5. 90 (x) 3 (x) 5.27 (=) 1422.9
6. 90 (x) 3 (=) + 270 *RULE OF SIGNS APPLIES*
7. 90 (x) 3 (=) - 270 *RULE OF SIGNS APPLIES*
8. .245 (x) .78 (x) 1.35 (=) .257985
9. 30 (x) 8 (x) 4 (x) 20 (=) -19,200 *RULE OF SIGNS APPLIES*
10. 120 (÷) 4 (=) 30
11. -426 (÷) 4 (=) -106.5
12. 520 (x) 123 (÷) 43560 (=) 1.468
13. .333 (x) 49 (x) 87 (÷) 27 (=) 52.577
14. 3.14 (x) 90 (x) 50 (÷) 37 (÷) 10 (=) 38.189
15. 8 (+) 27 (x) 60 (x) .0175 (÷) 6 (÷) 27 (÷) .10 (=) 2.269
16. 4 (÷) 12 (+) 2 (=) 2.333

    3 (÷) 12 (+) 25 (=) 25.250

    11 (÷) 12 (=) .916

    1 (÷) 12 (+) 5 (=) 5.083
17. 9" (÷) 12" (+) 6' (=) (6.750) 2" (÷) 12" (+) 112' (x) (6.750) (=) 757.127
18. 3 (÷) 12 (+) 16 = (16.250) 6 (÷) 12 (=) .5 (+) 8 (=) (8.500)

    4 (÷) 12 (+) 3 (=) 3.333 (x) (16.250) (x) (8.500) (=) 460.371 ft

NOTE: The ellipse ( ) indicates a manual memory operation. The calculator will not store memory in the division or multiplication function.

QUIZ NO. 7 Small System Compressor Trouble Shooting

1-E	3-B	5-B	7-B	9-E
2-D	4-D	6-E	8-E	

QUIZ NO. 8 Small System Compressor Trouble Shooting

1-A	3-D	5-A	7-A	9-A
2-E	4-C	6-B	8-E	

QUIZ NO. 9 Service and Maintenance

Section No. 1	Section No. 2	Section No. 3	Section No. 4	Section No. 5
1-A	11-D	20-C	32-B	41-D
2-A	12-C	21-B	33-A	42-B
3-A	13-A	22-A	34-D	43-D
4-A	14-C	23-D	35-B	44-C
5-D	15-A	24-A	36-B	45-C
6-A	16-A	25-C	37-B	46-D
7-A	17-C	26-C	38-C	47-D
8-A	18-C	27-A	39-B	48-C
9-C	19-D	28-C	40-B	49-C
10-B		29-B		50-A
		30-A		
		31-A		

QUIZ NO. 10 Service and Maintenance

Section No. 1	Section No. 2	Section No. 3	Section No. 4	Section No. 5
1-D	12-C	21-C	32-C	41-D
2-C	13-A	22-C	33-C	42-D
3-C	14-C	23-B	34-D	43-A
4-B	15-B	24-A	35-D	44-A
5-A	16-B	25-D	36-B	45-C
6-D	17-B	26-B	37-D	46-B
7-C	18-D	27-D	38-D	47-B
8-B	19-C	28-A	39-A	48-A
9-B	20-D	29-D	40-A	49-B
10-A		30-C		50-D
11-B		31-A		51-A
				52-C

QUIZ NO. 11 Gas Laws

1. Substituting for the formula,

$$V_2 = V_1 \times \frac{P_1}{P_2}$$

   a. The initial gas in the cylinder is at atmospheric pressure; therefore, P = 14.7 psia.

   b. The final pressure is 13 psig + 14.7 = 27.7 psia, P_2

   c. The initial volume is 3 cu ft, V_1

   $3\ ft^3 \times \frac{14.7}{27.7} = 1.59\ ft^3$, final volume, V_2 ANSWER A

2. Substituting for the formula.

$$P_2 = P_1 \times \frac{T_2}{T_1}$$

   a. The intitial pressure is 300 psig + 14.7 = 314.7 psia, P_1

   b. The final temperature is 60F + 460 = 520 deg absolute

   c. The initial temperature is 100F + 460 = 560 deg absolute

   $314.7 \times \frac{520}{560} = 292$ psia, final pressure, P_2 , but that answer does not appear in the answer band, then

   292, psia - 14.7 = 277.3 psig, ANSWER E

Note: In problems of these kinds, a practical approach is to round off the 14.7 to 15 and avoid the decimal.

QUIZ NO. 12 Temperature Conversion

1-B	3-D	5-A	7-D	10-D
2-C	4-D	6-B	8-B	11-A
			9-C	12-A

QUIZ NO. 13 Pipefitting

Section No. 1	Section No. 2	Section No. 3	Section No. 4
1-A	12-B	22-D	29-D
2-D	13-C	23-D	30-A
3-B	14-A	24-B	31-A
4-B	15-B	25-B	32-C
5-C	16-B	26-C	33-B
6-D	17-B	27-B	34-C
7-A	18-C	28-D	35-C
8-C	19-A		36-C
9-B	20-B		37-D
10-C	21-A		
11-D			

QUIZ NO. 14 Refrigeration Cycle

15 Evaporator
1 Expansion valve
7 Evaporator pressure regulator
8 Hi-lo cut-out
2 Solenoid valve
14 Liquid receiver
6 Sight glass
9 Oil separator
3 External equalizer
13 Relief Valve
4 Strainer
12 Vent valve
11 Condenser
5 Deydrator
10 Muffler

QUIZ NO. 15 Pulley Laws

1-D 5000/2500 x 10 = 20

2-A 12/3 x 1250 = 5000

3-A 1000/5000 x 10 = 2

4-B 4/30 x 3750 = 500

5-A 20/4 x 750 = 3750

QUIZ NO. 16 Fan Laws

1-B $\left(\frac{800}{1000}\right) \times 2000 \text{ cfm} = 1600 \text{ cfm}$

2-B $\left(\frac{800}{1000}\right)^2 \times 1.75 \text{ sp} = 1.12 \text{ sp}_2$

3-C $\left(\frac{800}{1000}\right)^3 \times 2 \text{ hp} = 1.024 \text{ hp}_2$

4-A 4000 cfm/80% = 5000 cfm

5000/4000 x 1000 rpm = 1250 rpm_2

5000/4000 x .5 sp = .78 sp_2

Quiz No. 17 Things to Remember

1. D	4. B	7. C
2. C	5. B	8. D
3. A	6. A	9. D
		10. B

Quiz No. 18 Formulas

1. D	4. D	7. C
2. A	5. D	8. B**
3. B	6. D*	9. A
		10. B

*The formula for this offset is A = C + (D x 1.414) C = 15 in. and D = 8 in.; therefore, 15 + (8 x 1.414) = 26.312 (10/32 = .312)

** $\frac{30 \times 6}{144 \text{ in.}} \times .90 = 1.125$, then, 1.125 x 600 cfm = 675 cfm

Quiz No. 19 Codes

1. D ASHRAE 15-89 § 7.4.2.2	6. A ASHRAE 15-89 § 12.9
2. A ASHRAE 15-89 § 5	7. D 75' x 100' x 10' = 75,000 ft³ 75,000 ft³ / 3 min = 25,000 cfm
3. B ASHRAE 15-89 § 5	8. B SFBC § 4606.7 (a)
4. D ASHRAE 15-89 § 5	9. D SFBC § 4606.7
5. D ASHRAE 15-89 § 10.12.1	10. D SFBC § 4103.1 (f)

ANSWERS TO WARM-UP QUIZ NO. 20

1. C See page 112

2. D A Btu is the amount of heat necessary to raise the temperature of a 1 lb of water, one degree F. Therefore, 35 lb of water x (70°F-60°F) = 350 Btu.

3. B Modern Refrigeration and Air Conditioning, page 32

4. B (Absorption = Rejection)

5. B See page 117 and 134, R-12 @ 100°F = 117.2

6. A Air Conditioning & Mechanical Trades, p 189 and 190: 3 gpm per ton x 5 ton = 15 gpm. See also Trane AC Manual, p 168. 14,400 Btuh or 240 Btum is most practical for comfort cooling for condesner water;

   Therefore;
   $$\frac{14{,}000 \text{ Btuh} \times 5 \text{ tons}}{500 \times 10} = 14 \text{ gpm}$$

7. C ASHRAE 15-1989, 10.12.1

8. D ASHRAE 15-1989, 7.6.4

9. B See "Standard Air", Trane AC Manual, page, 56.

10. B See OSHA definitions

ANSWERS TO FINAL EXAMS

All references are given in page numbers unless stated as "Section." Reference symbols are:

MR	Modern Refrigeration and Air Conditioning
TR	Trane Reciprocating Refrigeration
TAC	Trane Air Conditioning Manual
SMC	Standard Mechanical Code
SFBC	South Florida Building Code
SMACNA-F	SMACNA, Fiberglass Manual
SMACNA-L	SMACNA, Low Pressure Ducts
ASHRAE	ASHRAE Standard 15-1989
NFPA	National Fire Protection pamphlets

References for First Aid are from the American Red Cross pamphlets--check your local Red Cross.

MORNING SESSION

Part 1 Duct Symbols

A. 14 B. 9 C. 6 D. 13 E. 16 F. 10
G. 2 H. 17 I. 19 J. 20

Part 2 Field Wiring Symbols

A. 9 B. 6 C. 11 D. 12 E. 14 F. 2
G. 16 H. 10 I. 1 J. 17

Part 3 Wiring Diagram (See separate sheet following)

Part 4 Code Problems (Reference given in section numbers)

1. False, SFBC 4606.7, SMC 603.2
2. True, SFBC 4802.1: 10 x 14 x 8/7.5 = 149.33 cfm
3. True, SFBC 4901.1 (b) 4901.2 (a)
4. False, SFBC 4903.7 (b) 4610.1 (b)
5. True, SFBC 4103.2 (b)
6. False, SFBC 4103.3 (b)(2) and Vol. II; p. C-475. This has been changed from 12" to 6" to conform with NFPA 96.
7. True, SFBC 4802.1 (f) (1)
8. False, ASHRAE 15-89, 10.12.1
9. False, NFPA 96-22.1.2, SFBC 4103.3 (b)(1)
10. False, ASHRAE 15-89 9.1

Part 5 Refrigerating Systems Classification

See separate sheet following. The reference for these kinds of problems is ASHRAE 15-89.

ANSWER SHEET
WIRING DIAGRAM PROBLEM

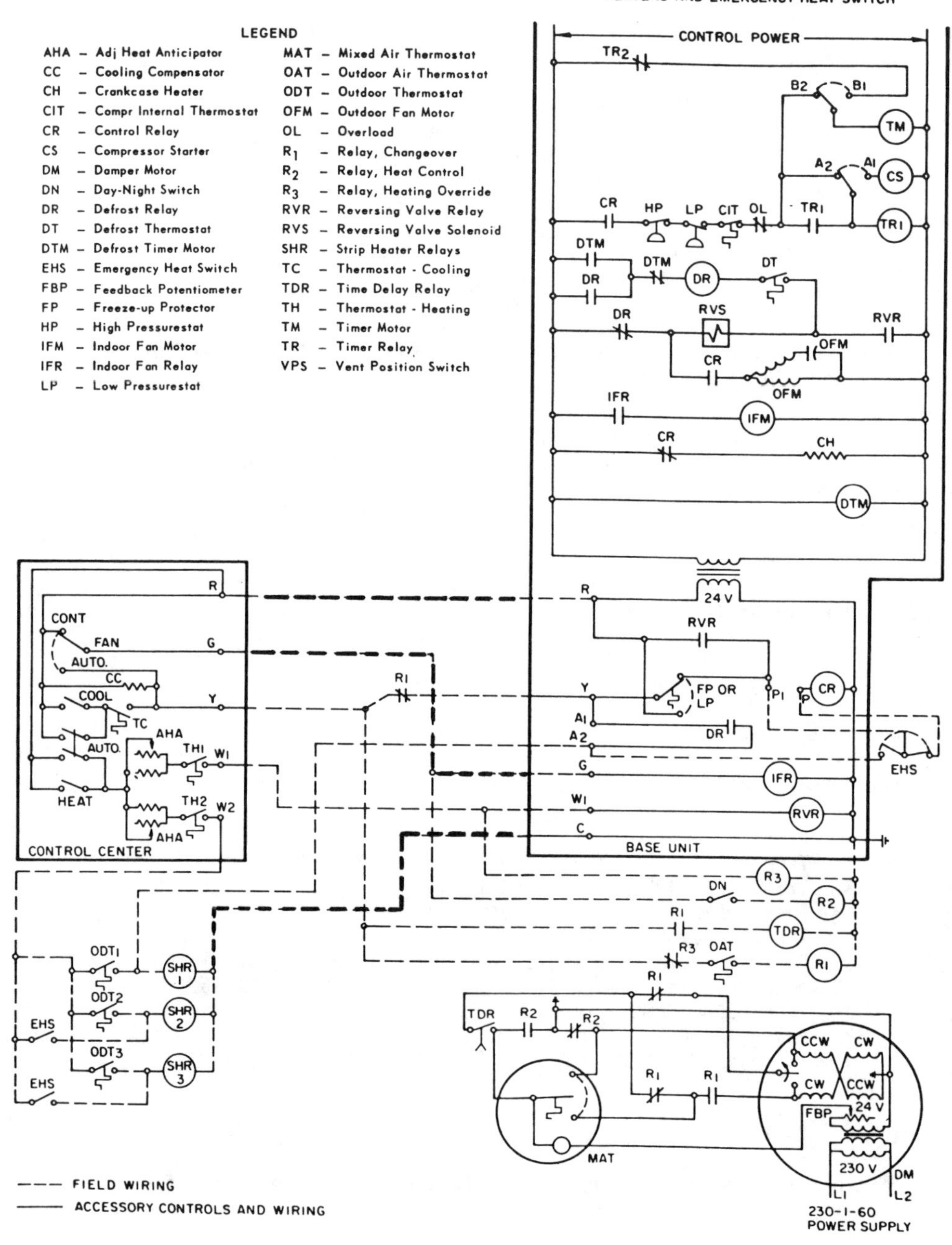

ANSWER SHEET FOR FIGURE 4

REFRIGERATING SYSTEM CLASSIFICATION PROBLEM

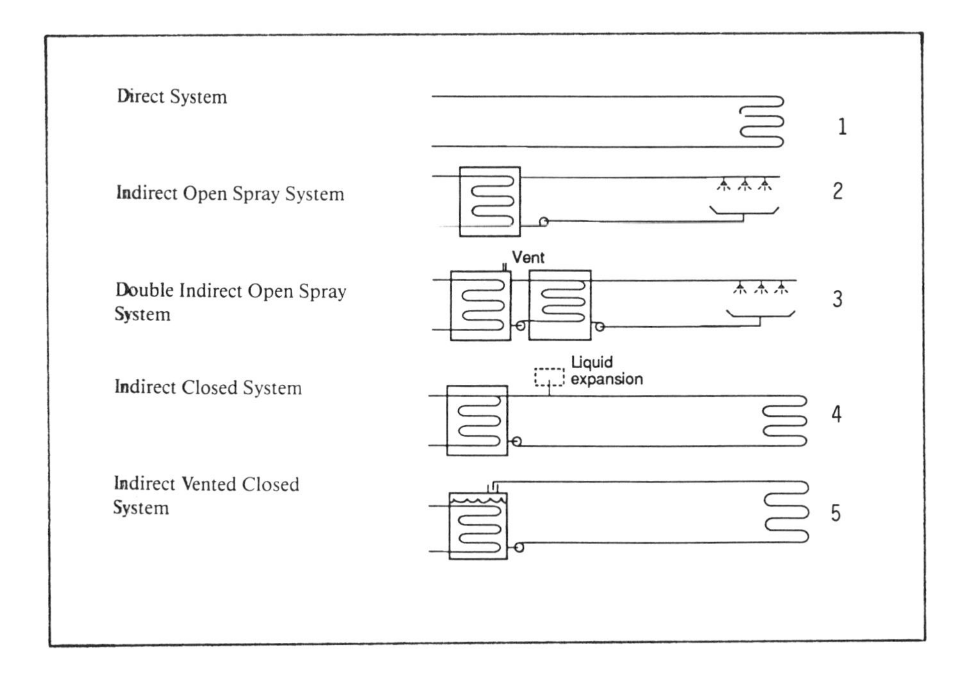

AFTERNOON SESSION

Part 1 First Aid--10 points, 2 points per question

1. F
2. T
3. F
4. T
5. T

Part 2 General--50 points, 1 point per question.

6. T MR 498
7. F MR 499
8. C MR 109
9. C MR 432, TR 79
10. D TR 78
11. A TR 53
12. D MR 63
13. C MR 364
14. C 384 (TR gives 12 hrs)
15. D MR 193,211
16. F
17. C MR 384 (or 500 microns)
18. A SMACNA-L, A-1
19. B MR 260
20. D (by calculation)
21. B MR 232
22. A
23. D MR 206
24. B SMACNA-F, 3-4
25. D SMACNA-F, 5-2
26. B SMACNA-F, 2-3
27. B SMACNA-F, 2-3
28. B MR 361
29. D ASHRAE 18, TR 118, MR 361
30. A MR 361
31. C MR 949 (1 ft WG x 62.32 =p
62.32/12 in. = 5.1
32. A MR 794
33. A MR 49
34. B TR 85
35. C MR 150
36. D MR 65
37. A TR 9
38. D MR 212
39. D MR 432
40. A TR 53
41. C MR 506
42. C MR 361
43. B MR 361
44. F ASHRAE 15-89, 10.12.1
45. D ASHRAE 15-89, Table 1
46. C ASHRAE 15-89, Table 1
47. D ASHRAE 15-89, Fig. 1
48. C ASHRAE 15-89, 7.1
49. C
50. B ASHRAE 15-89, 7.5.3
51. F SFBC , § 406.7 (h) (1) (1)
52. T SMC § 507
53. D ASHRAE 15-89, 12.4
54. T ASHRAE 15-89, 12.9
55. T NFPA 90A Sect. 3-1.1

Part 3 First Aid -- 10 points, 2 points per question

1. C
2. A
3. T
4. F
5. T

Part 4 Steam Specialties--30 points, 3 points per question.

1. C
2. K
3. C
4. B
5. A
6. B
7. A
8. E
9. A
10. A
11. H
12. D
13. J
14. J
15. F

PART 6

APPENDIX

TABLE A.1

USEFUL CONVERSION FACTORS		
Multiply	By	To Obtain
Atmosphere	29.92	Inches of mercury
Atmosphere	33.93	Feet of water
Atmosphere	14.70	Pounds per sq in.
Atmosphere	1.058	Tons per square foot
Barrels (oil)	42	Gallons
Boiler horsepower	33,475	Btu per hour
Boiler horsepower	34.5	Pounds water evaporated from and at 212F
Btu	778	Foot-pounds
Btu	0.000393	Horsepower-hours
Btu	0.000293	Kilowatt-hours
Btu	0.0010307	Pounds water evaporated from and at 212F
Btu per 24 hr	0.00000347	Tons of refrigeration
Btu per hour	0.00002986	Boiler horsepower
Btu per hour	0.000393	Horsepower
Btu per hour	0.000293	Kilowatts
Btu per inch per sq ft per hr per F.	0.0833	Btu per foot per sq ft per hour per F
Cubic feet	1,728	Cubic inches
Cubic feet	7.48052	Gallons
Cubic feet of water	62.37	Pounds (at 60F)
Cubic feet per minute	0.1247	Gallons per second
Feet of water	0.881	Inches of mercury (at 32F)
Feet of water	62.37	Pounds per sq ft
Feet of water	0.4335	Pounds per sq in
Feet of water	0.02950	Atmospheres
Feet per minute	0.01136	Miles per hour
Feet per minute	0.01667	Feet per second
Foot-pounds	0.001286	Btu
Gallons (U.S.)	0.1337	Cubic feet
Gallons (U.S.)	231	Cubic inches
Gallons of water	8.3453	Pounds of water (at 60F)
Horsepower	550	Foot-pounds per sec
Horsepower	33,000	Foot-pounds per min
Horsepower	2,546	Btu per hour
Horsepower	42.42	Btu per minute
Horsepower	0.7457	Kilowatts
Horsepower (boiler)	33,475	Btu per hour

TABLE A.1 (Continued)

Multiply	By	To Obtain
Inches of mercury (at 62F)	13.57	In. of Water (at 62F)
Inches of mercury (at 62F)	1.131	Ft of water (at 62F)
Inches of mercury (at 62F)	70.73	Pounds per sq ft
Inches of mercury (at 62F)	0.4912	Pounds per sq in
Inches of water (at 62F)	0.07355	Inches of mercury
Inches of water (at 62F)	0.03613	Pounds per sq in
Inches of water (at 62F)	5.202	Pounds per sq ft
Inches of water (at 62F)	0.002458	Atmospheres
Kilowatts	56.92	Btu per minute
Kilowatts	1.341	Horsepower
Kilowatt-hours	3415	Btu
Latent heat of ice	143.33	Btu per pound
Pounds	7,000	Grains
Pounds of water (at 60F)	0.01602	Cubic feet
Pounds of water (at 60F)	27.68	Cubic inches
Pounds of water (at 60F)	0.1198	Gallons
Pounds of water evaporated from and at 212F	0.284	Kilowatt-hours
Pounds of water evaporated from and at 212F	0.381	Horsepower-hours
Pounds of water evaporated from and at 212F	970.4	Btu
Pounds per square inch	2.0416	In. of mercury
Pounds per square inch	2.309	Ft of water (at 62F)

TABLE A.3

DECIMALS OF A FOOT

In Frac.	0"	1"	2"	3"	4"	5"	6"	7"	8"	9"	10"	11"
0	.000	.083	.167	.250	.333	.417	.500	.583	.667	.750	.833	.917
1/8"	.010	.094	.177	.260	.344	.427	.510	.594	.677	.760	.844	.927
1/4"	.021	.104	.188	.271	.354	.438	.521	.604	.688	.771	.854	.938
3/8"	.031	.115	.198	.281	.365	.448	.531	.615	.698	.781	.865	.948
1/2"	.042	.125	.208	.292	.375	.458	.542	.625	.708	.792	.875	.958
5/8"	.052	.135	.219	.302	.385	.469	.552	.635	.719	.802	.885	.969
3/4"	.063	.146	.229	.313	.396	.479	.563	.646	.729	.813	.896	.979
7/8"	.073	.156	.240	.323	.406	.490	.573	.656	.740	.823	.906	.990

To change decimals of a foot to inches, multiply the decimal by 12.

To change inches to decimals of a foot, divide inches by 12.

TABLE A.3

INCH — FOOT — DECIMAL CONVERSION

Inches, Fractions	Inches, Decimals	Feet, Decimals	Inches, Fractions	Inches, Decimals	Feet , Decimals	Inches, Fractions	Inches, Decimals	Feet, Decimals
1/64	.0156	.0013	11/32	.3438	.0287	43/64	.6719	.0560
1/32	.0313	.0026	23/64	.3594	.0299	11/16	.6875	.0573
3/64	.0469	.0039				45/64	.7031	.0586
1/16	.0625	.0052	3/8	.3750	.0313	23/32	.7188	.0599
5/64	.0781	.0065	25/64	.3906	.0326	47/64	.7344	.0612
3/32	.0938	.0078	13/32	.4063	.0339			
7/64	.1094	.0091	27/64	.4219	.0352	3/4	.7500	.0625
			7/16	.4375	.0365	49/64	.7656	.0638
1/8	.1250	.0104	29/64	.4531	.0378	25/32	.7813	.0651
9/64	.1406	.0117	15/32	.4688	.0391	51/64	.7969	.0664
5/32	.1563	.0130	31/64	.4844	.0404	13/16	.8125	.0677
11/64	.1719	.0143				53/64	.8281	.0690
3/16	.1875	.0156	1/2	.5000	.0417	27/32	.8437	.0703
13/64	.2031	.0169	33/64	.5156	.0430	55/64	.8594	.0716
7/32	.2188	.0182	17/32	.5313	.0443			
15/64	.2343	.0195	35/64	.5469	.0456	7/8	.8750	.0729
			9/16	.5625	.0469	57/64	.8906	.0742
1/4	.2500	.0208	37/64	.5781	.0482	29/32	.9063	.0755
17/64	.2656	.0221	19/32	.5938	.0495	59/64	.9219	.0768
9/32	.2813	.0234	39/64	.6094	.0508	15/16	.9375	.0781
19/64	.2969	.0247	5/8	.6250	.0521	61/64	.9531	.0794
5/16	.3125	.0260	41/64	.6406	.0534	31/32	.9688	.0807
21/64	.3281	.0273	21/32	.6563	.0547	63/64	.9844	.0820

American Standard

Scheme for the Identification of Piping Systems

Object and Scope

1 The scheme is intended to establish a common code to assist in the identification of materials conveyed in piping systems and is intended to form an acceptable basis for a universal scheme. The use of this standard will promote greater safety and will lessen the chances of error, confusion or inaction.

2 This scheme concerns only the identification of piping systems in industrial and power plants. It does not cover pipes buried in the ground or electrical conduits.

Definitions

3 **Piping Systems.** For the purpose of this scheme, piping systems shall include in addition to pipes of any kind: fittings, valves and pipe coverings. Supports, brackets, or other accessories are specifically excluded from applications of this standard. Pipes are defined as conduits for the transport of gases, liquids, semi-liquids or plastics, but not solids carried in air or gas.

4 **Fire Protection, Materials and Equipment.** This classification includes sprinkler systems and other fire-fighting or fire protection equipment. The identification for this group of materials may also be used to identify or locate such equipment as alarm boxes, extinguishers, fire blankets, fire doors, hose connections, hydrants and any other fire-fighting equipment.

5 **Dangerous Materials.** This group includes materials which are hazardous to life or property because they are easily ignited, toxic, corrosive at high temperatures and pressures, productive of poisonous gases or are in themselves poisonous. It also includes materials that are known ordinarily as fire producers or explosives.

6 **Safe Materials.** This group includes those materials involving little or no hazard to life or property in their handling. This classification includes materials at low pressures and temperatures, which are neither toxic nor poisonous and will not produce fires or explosions. People working on piping systems, carrying these materials run little risks even though the system had not been emptied.

7 **Protective Materials.** This group includes materials which are piped through plants for the express purpose of being available to prevent or minimize the hazard of the dangerous materials above mentioned. It would include certain special gases which are antidotes, to counteract poisonous fumes, piped for the express purpose of release in case of danger. This classification also covers protective materials for purposes other than for fire protection which is covered under Section 4.

Method of Identification

8 Positive identification of a piping system content shall be by lettered legend giving the name of the content in full or abbreviated form. Arrows may be used to indicate the direction of flow. Where it is desirable or necessary to give supplementary information such as hazard or use of the piping system content, this may be done by additional legend or by color applied to the entire piping system or as colored bands. Legends may be placed on colored bands.

Examples of legend to give both positive identification and supplementary information as regards hazards or use are:

Water	– Fire protection
Ammonia	– Anhydrous – Dangerous liquid and gas
Acetone	– Extremely flammable liquid
Hydrogen	– Extremely flammable gas
Air	– High-pressure gas
Carbon Dioxide	– Fire protection

NOTE: Manual L-1, third revision, 1953, published by Manufacturing Chemists Association, Inc., is a valuable guide in respect to supplementary legend.

9 When color, applied to the entire piping system or as colored bands, is used to give supplementary information it shall conform to the following:

CLASSIFICATION	PREDOMINANT COLOR
F – Fire-protection equipment	Red
D – Dangerous materials	Yellow(or orange)
S – Safe materials	Green (or the achromatic colors, white, black, gray or aluminum)
and, when required,	
P – Protective materials	Bright blue

Extracted from American Standard *Scheme for the Identification of Piping Systems* (ANSI A13.1-1956), with permission of the publisher, The American Society of Mechanical Engineers, 345 East 47th St., New York, N.Y.

10 The above colors have been chosen to identify the main classifications because they are readily distinguishable one from another under normal conditions of illumination. Fig. 1 shows the four main classification color bands, color of the legend letters and their suggested placement location, also the recommended width of color band "A" together with the size of the legend letters "B" for various pipe diameters.

11 Color bands, if used, shall be painted or applied on the pipes to designate to which of the four main classifications its contents belongs. The bands should be installed at frequent intervals on straight pipe runs (sufficient to clearly identify) close to all valves, and adjacent to all change-in-directions, or where pipes pass through walls or floors. The color identification may be accomplished by the use of decals or plastic bands which are made to conform with the standards. If desired the entire length of the piping system may be painted the main classification color.

12 Attention has been given to visibility with reference to pipe markings. Where pipe lines are located some distance above the normal line of the operator's vision the lettering should be placed below the horizontal center line of the pipe as shown in Fig. 1.

13 In certain types of plants it may be desirable to label the pipes at junction points or points of distribution only while at other locations the markings may be installed at necessary intervals all along the piping, close to valves and adjacent to change-in-directions. In any case the number and location of identification markers should be based on judgment for each particular system of piping.

14 Regarding the type and size of letters, the use of stencils of standard sizes ranging in height from ½ in. to 3½ in. is recommended. For identification of pipe less than ¾ in. in diameter the use of a tag is recommended. The lettering or the background may be of the standard color. (See Fig. 1).

15 In cases where it is decided to paint the entire piping, the color and sizes of legend letters stencilled on the piping for identification of material conveyed should conform to the specifications shown in Fig. 1.

Key to Classification Color of Bands-
Color of Legend Letters-
Legend Placement-Width of Color Bands and Size of Letters for Various Diameter Pipes

KEY TO CLASSIFICATION OF PREDOMINANT COLORS FOR BANDS		COLOR OF LETTERS FOR LEGENDS
F – Fire protection	Red	White
D – Dangerous	Yellow	Black
S – Safe	Green	Black
P – Protective	Blue	White

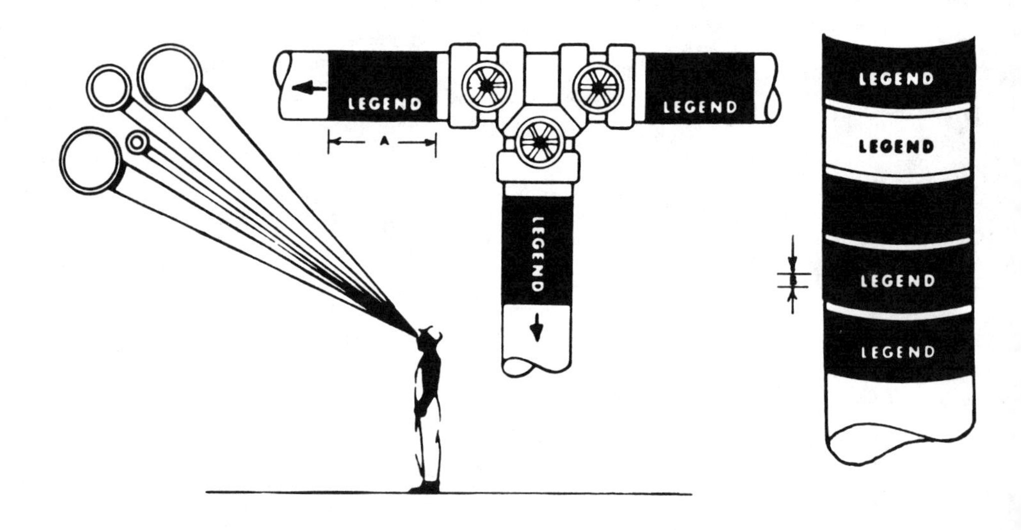

Outside Diameter of Pipe or Covering	Width of Color Band A	Size of Legend Letters B
¾ to 1 ¼	8	½
1 ½ to 2	8	¾
2 ½ to 6	12	1 ¼
8 to 10	24	2 ½
Over 10	32	3 ½

All dimensions are given in inches.

FIGURE 6.2

INDEX